A Surveyor's Guide to the RICS HomeBuyer Service

Phil Parnham

Published by the Royal Institution of Chartered Surveyors
Surveyor Court
Westwood Business Park
Coventry CV4 8JE
UK

www.ricsbooks.com

No responsibility for loss or damage caused to any person acting or refraining from action as a result of the material included in this publication can be accepted by the author or RICS.

ISBN 978 184219 511 6

© Royal Institution of Chartered Surveyors (RICS) October 2009. Copyright in all or part of this publication rests with RICS, and save by prior consent of RICS, no part or parts shall be reproduced by any means electronic, mechanical, photocopying or otherwise, now known or to be devised.

Typeset and printed by Page Bros, Norwich

Contents

Foreword	**vi**
About the author	vi
Introduction	**1**
Who this book is for	1
What the Guide does	1
What the Guide does not do	1
1. The New HomeBuyer Report	**3**
The need for change	3
The HomeBuyer Report (3rd Edition)	4
The new practice notes	4
Additional advice – 'up sell' your services	6
Before you begin	6
2. Preparing for the Inspection	**7**
Knowing your limitations	7
Pre-inspection checks	8
Environment Agency	8
Defra noise maps	9
Geological information	9
Health Protection Agency	11
OFCOM (mobile phone base stations)	11
Free environmental risk assessment services	11
Regional construction techniques and local knowledge	12
Listed buildings and conservation areas	12
Tools and equipment	12
Binoculars	13
Ladder	13
Measuring devices	13
Electronic moisture meter	13
Torch	13
Lifting equipment/crow bar	13
Spirit level	14
Pocket mirror	14
Plumb bob	14
Compass	14
Digital camera	14
Protective clothing	14
Health and safety equipment	14
Health and safety matters	15
The legislation	15
Risk assessments	19
Property-specific safety issues	22
Liaising with the occupiers	24
Using seller questionnaires	25
Building a relationship with the seller – on arrival	25
Dealing with seller enquiries	25

3.	**The Inspection Procedure**	**27**
	How far do we go?	27
	Inspection of flats	30
	Inspection procedure	32
	Site notes	32
4.	**Understanding the Condition Ratings**	**37**
	Why have condition ratings been introduced?	37
	Deconstructing the definitions	37
	Defect	38
	Normal maintenance	39
	Serious defects	39
	Urgent defects	40
	Further investigations	40
	Building services: the special status of further investigations, 'not inspected' and condition rating	41
5.	**Developing a Consistent Approach – the Condition Rating Protocol**	**45**
	The Condition Rating Protocol	46
	Broader issues that affect condition rating decisions	46
	Condition rating case studies	46
	Condition rating of roof coverings (E2: Roof coverings)	47
	Condition rating of heating systems (G4: Heating)	53
	Health and safety issues and the condition rating system	59
	An objective method for assessing health and safety hazards	60
	Applying the condition rating to multiple building elements	65
	Condition rating of miscellaneous elements	66
	Decorations	66
	Conservatory and porches – the new building element E7	66
	Built in elements	67
	Insulation, ventilation and condensation	68
	Garages, outbuildings and grounds	70
	Condition rating of 'risks to the building' (section J1)	75
	Factors that can affect the value of the property but are not condition rated	75
	Market value and condition rating	76
	Condition rating building elements in flats	77
	Elements that are exclusive to the inspected flat	77
	Elements that are shared with the block	77
	Services	78
	External shared areas	79
6.	**Writing the Report**	**81**
	How to complete the comment boxes	83
	Descriptive section	83
	Advisory section	83
	Other report-writing issues	84
	How long should the report be?	84
	Reporting on further investigations	84
	Ordering the text	84
	Cosmetic matters and minor defects	84
	Report-writing examples	85
	Completing section I: Issues for your legal advisers	86
	Guarantees and warranties	87
	Overall opinion	88

7.	**Making the HomeBuyer Work**	**89**
	Expanding the market share	89
	Marketing the HBR	89
	Support from RICS	90
	Building for the future	91

Appendices   93

Appendix 1		**95**
	(a) HBR site notes	95
	(b) HBR site notes (room-by-room supplement)	121
	(c) HBR site notes (for dictation)	147
Appendix 2	Sellers' questionnaire	161
Appendix 3	Market valuation analysis	165
Appendix 4	HBR site notes (sample)	167
Appendix 5	Completed report (sample)	193

Index   227

Foreword

Packed full of illustrated examples, together with a sample completed report, this book gives the surveyor a practical, balanced, and sensible approach to producing the new RICS HomeBuyer Report.

The book explains just how the new RICS HomeBuyer Service requires us to change the way we write our reports. Our thoughts and our reports will be organised according to the building elements. Phil follows the RICS Practice Note, and offers us a clear and logical strategy for the application of condition ratings to the building elements, and shows us how to manage sub-elements. The way he deals with services, and the assessment of risks and hazards, aims to result in effective decision making.

Particularly useful for us all is the review of health and safety issues – not only as required by the report as part of the assessment of the property, but also to ensure that we get safely from the office to the property and back. There is also advice for dealing with the anxious owner-occupier in whose property we carry out our work.

The sections on flats are welcome for the approach to dealing with the common elements and the inspection of common parts, services and grounds.

Written by an experienced surveyor with a 'weather eye' on the increasingly litigious climate, this book contains plenty of advice and examples on the use of site notes to ensure that an evidence base exists to support the final report.

This book will help us all – however experienced we are in surveying residential property – to deal with the change in approach required by the new HomeBuyer Service, and to write our reports more clearly for the benefit of our clients.

Barry K Hall FRICS MCIArb
(Chair of the RICS Residential Survey &
Valuation Group)
September 2009

DEDICATION

To the woman who introduced me to curried eggs.

ABOUT THE AUTHOR

Phil Parnham is a chartered building surveyor and professional development director at BlueBox partners. He has many years' experience in higher education where he developed a range of training services for those working in the residential property sector. At BlueBox, he continues to develop this work as well as launching a number of unique and innovative training initiatives, including distance learning and assessment programmes for those working in the property sector.

Phil has also written the *RICS Survey Writer Sample Phrases for the RICS HomeBuyer Report*, and co-written a number of other books including *The Home Inspector's Handbook* and *Domestic Energy Assessor's Handbook* for RICS, and *Residential Property Appraisal* for Spon Press.

Introduction

As the name suggests, this book is a surveyor's guide. It has been written to help surveyors get to grips with the new RICS HomeBuyer Report 2009 (HBR), which was launched by the Royal Institution of Chartered Surveyors (RICS) in July 2009.

This book is not an extension of the RICS practice notes (*The RICS HomeBuyer Service 2009*), but a publication designed to support residential practitioners tackle the problems and issues they encounter everyday. The advice in this book is the result of the feedback and contributions made by the hundreds of surveyors I have met and worked with during conferences and training workshops.

It is a practitioner-focused book that is not meant to sit on your bookshelf; but to go with you on the survey.

The book also aims to help you adjust to the new culture ushered in by the HBR 2009. The use of condition ratings and a simpler, more direct style of reporting make the third edition of this familiar product completely different to its predecessors. It challenges experienced practitioners to review and reflect on how they assess residential properties, and the style and content of their reports.

The credit crunch and the recession have changed the residential sector forever. Our clients expect something different, and the fresh approach offered by the HBR gives all of us a real opportunity to market our services directly to the public. If we are to be less dependent on third parties for our work in the future, we need a new range of products to work with. Thanks to the far-sighted Residential Practice Team at RICS, the HBR 2009 should be the first in series of services designed to provide surveyors with a flexible response to these challenging times. I hope this book makes a contribution to this initiative.

WHO THIS BOOK IS FOR

The HomeBuyer Report 2009 is a licensed RICS product that only chartered surveyors can produce. Consequently, this book is designed to suit:

- experienced practitioners who are familiar with the RICS HomeBuyer Survey & Valuation Service 2005 (second edition); and
- new entrants to the residential sector who are working towards membership of RICS.

The book will also suit those experienced surveyors who are changing the focus of the services they offer – building surveyors who want to develop a 'mid-range' product, or valuation surveyors who want to add condition assessment to their profile, for instance.

Other professionals and commentators may also find the book of use, because it deconstructs the HBR into its constituent parts of desk-top preparation, inspection, allocation of condition ratings and reporting.

WHAT THE BOOK DOES

This book is not an official RICS publication. Although I have worked closely with a number of officials, the focus of the publication is on the needs of the practitioner at the 'coal face'. I have used the practice notes to the 3rd Edition (*The RICS HomeBuyer Service 2009*) as a starting point to:

- produce protocols and methodologies that will support your decision making;
- illustrate these techniques by using realistic case studies; and
- offer a decision-making framework that will help you produce balanced and consistent decisions, particularly where there are gaps in the practice notes, until the official guidance is amended in the future.

Throughout this book, direct references to particular parts of the practice notes are shown in square brackets, as follows: [paragraph 1.4, page 3].

WHAT THE BOOK DOES NOT DO

This book does not go back to first principles. It assumes you have a sound knowledge of residential property, especially:

- residential construction;
- building pathology;
- building services;
- valuation issues; and
- regulatory and legal issues relevant to the residential sector.

Working closely with experienced practitioners during my training seminars and workshops, it has become clear to me that many have not kept pace with the changes affecting the residential sector. There are particular knowledge gaps in regulatory changes (for example, building regulations, environmental and safety regulations), technological developments (especially with building services and modern methods of construction) and the range of useful information sources available through the Internet. These new developments are not associated with the HBR, but are matters that surveyors should now be incorporating into their daily work. What the HBR does is provide the opportunity for experienced practitioners to upgrade and strengthen their knowledge base and, along the way, that will make this new product even more effective and attractive to the public.

This book touches on many of these issues but you may find that you have to go elsewhere to fill the gaps in your knowledge.

I hope you find this book useful. My previous publications have taught me a number of important lessons:

- because of the long lead-in times, any book is out of date as soon as it is published (I am typing this in May for a book that will be published in mid-October at the earliest);
- no matter how much I research topics, there will always be someone out there who knows far more than I do;
- there is bound to be a regulatory or legislative issue I have not included; and
- some readers simply will not agree with my view of the world.

1. The New HomeBuyer Report

THE NEED FOR CHANGE

The RICS Homebuyer Survey and Valuation Service (HSV), first introduced in 1993, aimed to be a 'concise' report, sitting in that difficult middle ground between the mortgage valuation and the more extensive building survey. Updated in 1997 and again in 2005, it included more information on the condition of the property than the mortgage valuation but did not go as far as the building survey. It was seen as an ideal product for surveyors who specialise in valuation of domestic properties.

Generally speaking, the format has been successful and has served both residential surveyors and the public well. However, the social, political and financial context has changed dramatically over the past 16 years, rendering the HSV somewhat tired and dated. The credit crunch and the subsequent recession have delivered the coup de grâce to the HSV. But the economic catastrophe cannot be blamed for everything. A number of other influences persuaded RICS to take a long hard look at the HSV:

- Both surveyors and clients were often confused by some of the definitions in the guidance. For example, 'urgent' and 'significant' were often confused. A defect could be:
  - urgent and significant;
  - urgent but not significant;
  - significant but not urgent.

  Additionally, some matters could be 'of concern' but not considered urgent or significant.
- The 'Action' paragraphs baffled many; surveyors were never quite sure of the differences between F1, F2 or F3. Where an F1 was tagged, many surveyors expressed it incorrectly.
- Although the 'overall opinion' was at the front of the report, section F was towards the end, making it difficult for the reader to find the summary. It made an assumption that the client had to read the report from front to back and only then could they get to the summary.

If surveyors were 'not quite sure', then it is not surprising that their customers shared the same confusion.

Other influences include:

- The launch of the Home Information Packs (HIPs), although much criticised, has introduced the concept of 'up front' property information.
- The Carsberg 'Review of Residential Property' in June 2008 made a number of far reaching recommendations, including making the HIPs voluntary and encouraging people involved in transactions to 'understand the various possibilities'. With an election likely in 2010, RICS needed to be ready to respond to different political influences.
- The Single Survey in Scotland, introduced by the Scottish Parliament, includes a mandatory condition report that uses condition ratings and can only be produced by chartered surveyors.
- Once the recession is over, lenders may make greater use of automated valuation methods. This could result in home buyers not even getting a mortgage valuation, and they may then want cost-effective reassurance to replace it.

In September 2007, RICS commissioned a market research project to see how the home-buying public viewed the various survey products. Over a two-month period, the researchers ran six consumer focus groups and an expert panel workshop across the UK. People were asked to state their views about home inspections and surveys. The objective was to gain insights into why only 20 per cent of UK home buyers have the property they are buying surveyed and how RICS could address this situation.

The main findings were:

- There was a lack of knowledge of the different type of 'surveys' and many people thought that valuation reports were 'proper surveys'.
- People had little concept of the role of the chartered surveyor and no understanding of RICS. The only exception was that a number of people recognised the 'lion' logo.
- A number felt that the HSV was too detailed and confusing, and preferred something with more 'ticked boxes'. They wanted it written in plain English.
- Many consumers thought that they could spot any defects themselves, while others preferred to take a tradesperson or builder to look around the home.

- Some felt that moving was an expensive business and finding £400–£500 was just too much additional expense. Others felt that when they had found a property that they liked, they were worried that the surveyor would spoil their dreams.

The same research team also interviewed a group of experience chartered surveyors on the same topic. The conclusions included:

- The current 'level two' product could be significantly improved by including a simple bullet point list at the front and some indication of the potential costs involved in remedying the defects.
- Simplified report writing could reduce cost.
- If the public realised that automated valuation models (AVMs) were being used more widely, they may be more interested in having a survey done.

This is a very brief summary of what is a long and detailed report, but nevertheless highlights the fact that the prevailing range of inspection and survey products did not meet customer needs. This set the background for a review of the existing HSV by a Product Review Group set up by RICS.

THE HOMEBUYER REPORT (3RD EDITION)

After several months of intensive development, the Product Review Group approved and published the practice notes (*RICS HomeBuyer Service 2009*) and launched the HomeBuyer Report in April 2009. The HBR 2009:

- has a crisp modern design that uses colour;
- repeatedly uses the RICS logo to promote the 'brand';
- uses condition ratings; and
- is supported by full practice notes.

Rather than reinvent the format completely, the Review Group called it the '3rd Edition' to emphasise that it remains the benchmark for 'level two' surveys. In my view, this is sensible. The residential sector is familiar with this concept and it is better to help the product adapt and change.

Although much of the 3rd Edition will be familiar, there are a number of important changes. Here are just a few of them:

- The Summary is at the front of the report, together with the 'overall opinion', so it is the first thing the reader sees.
- It is now an elemental report, so categories based on defects have gone (for example, the old C1–C4 are now reported under the building elements).
- There are new elements for 'water heating' and 'bathroom fittings' and 'garages' while 'decorations' have been subsumed within the elements of which they are part.

The new practice notes

The new practice notes (*The RICS HomeBuyer Service 2009 3rd Edition*) very clearly describe the nature and extent of the new service and I do not want to repeat the guidance here. Instead, I want to pick out a number of practical issues currently causing concern.

Figure 1.1 The new report front page (left) and the old report (HSV)

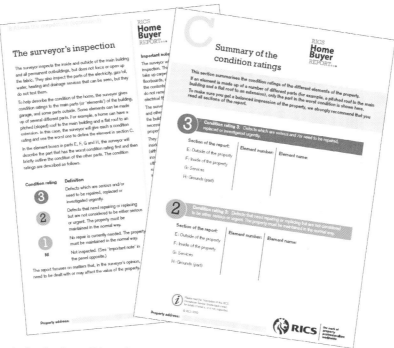

Figure 1.2 The new report style, showing the condition ratings

Relationship with the Home Information Pack

Although most surveyors would like to ignore it, for the time being the Home Information Pack (HIP) is part of our professional landscape. The practice notes state that the surveyor should '... obtain a copy of the Property Information Questionnaire (PIQ) and the Energy Performance Certificate (EPC), which are both required documents in the HIP' [paragraph 1.5, page 4]. The HIP has added a stage to the pre-inspection process and presents a number of challenges for the surveyor:

- Who holds the HIP? In most circumstances both the agent and the owner will have a copy and so viewing it should be straightforward. Occasionally, though, where the HIP has been assembled by an on-line supplier, or the property is not occupied, this might not be so easy. Also, some parties to the sale may be deliberately obstructive. Some see surveyors as deal breakers and will not go out of their way to help.
- Even when you get hold of the HIP, will it be any use? One surveyor recently told me he has never seen so many printed pages contain so little information. Even the property information questionnaire (PIQ) itself can be close to useless. Whoever completes it always has the option of entering 'don't know' against the questions.
- The energy performance certificate (EPC) is usually more straightforward. RICS has been very clear by saying that the energy efficiency rating and the environmental impact ratings are simply transposed into the HBR. The figures could be obtained from the property details if the HIP cannot be traced.

If it has been completed correctly, the PIQ can 'alert' you to problems at the property and play a useful role, but do not allow the HIP to distract you.

My main point is that you need to adjust your inspection procedure to allow for tracking down the HIP and studying the PIQ. Despite your reasonable efforts (always make a note of these in your site notes), where you do not manage to view the PIQ or the EPC, it must be reported in the HBR.

Communication with the client

The RICS research also revealed that many members of the public do not fully understand the differences between the survey and report types. The practice notes place a clear duty on the surveyor [paragraph 1.6, page 4] to check 'that the client:
(a) is making the appropriate choice of survey; and
(b) has a clear understanding of the key elements of the service.'

The practice notes suggest that this is done '... where possible through communication (for example, telephone conversation or email)'. The implication of this statement is that surveyors do not usually have a direct relationship with their clients (i.e. that secretaries and other office staff often explain what the service involves and then pass on basic information to the surveyor, usually just before the inspection). But as the practice notes now reinforce, establishing a direct relationship not only helps to identify what the client wants from the report, it also avoids misunderstandings that can often lead to complaints and sometimes court action.

Requirements of the HomeBuyer Service

Section 2.2 of the practice notes states: 'It is mandatory to use the specified format and wording, without variation. No departure from the specified mandatory elements of the service is permitted'.

Yet anecdotal information suggests that a number of practitioners vary the format, and many offer the old HSV without the valuation. Those who were prepared to discuss this matter gave the following reasons:

- Many clients have their valuation and finance already set up. All they want is an idea about the property's condition, but do not want to pay for a building survey.
- The surveyors concerned were not valuers but their clients wanted the HSV.
- Getting professional indemnity insurance for valuation work is expensive, but cheaper for condition reports and so they simply do not include a valuation as part of the services they offer to clients.

This clear breach of the practice notes indicates that both surveyors and their clients want more flexible products. Although RICS is working on this (see page 91), the current HBR comes as a package that **must** include a valuation and this cannot be varied in any way.

ADDITIONAL ADVICE – 'UP SELL' YOUR SERVICES

Although the HomeBuyer Service has its mandatory components, surveyors have always been able to offer other services through the 'Additional advice' provision. The practice notes [paragraph 2.5, page 6] give some typical examples:

- 'inspection of parts or areas not normally examined;
- general comment on plans for limited extensions to the property;
- more extensive inspection of a flat's shared areas or services … '

The main condition is that the additional advice must not change the nature of the service and must be reported in the HBR itself. This provision allows considerable flexibility to 'up-sell' a greater range of products. Admittedly, it is difficult to persuade clients to pay more in the current climate, but it does provide an opportunity to create a bespoke service for a reasonable fee. I will return to this theme in chapter 7.

BEFORE YOU BEGIN

This quick review of the history and context of the new HBR is designed to set the scene for the rest of this book. The following chapters aim to provide practical guidance that will suit both existing practitioners and new entrants. However, this is not a stand-alone publication and you should already have the following documents: *The RICS HomeBuyer Service 2009, 3rd Edition Practice Notes*, RICS 2009; *RICS HomeBuyer Report Property Survey and Valuation*, RICS 2009; and *The RICS Valuation Standards* (the Red Book) (see UKPS 4.1). As a member of RICS, you should be able to get the first two free from the RICS website.

2. Preparing for the Inspection

This chapter covers all matters relating to preparing to carry out the inspection for a HomeBuyer Report (HBR), including:

- the pre-inspection checks;
- inspection equipment;
- health and safety issues;
- liaison with the occupiers.

KNOWING YOUR LIMITATIONS

Before you accept an instruction to carry out an HBR, you must be sure that you are competent to assess the property. Unlike previous versions, the new practice notes [paragraph 2.3, page 5] are not prescriptive about the nature of the dwelling. For example, the practice notes say: 'The service applies to houses, bungalows and flats that are conventional in type and construction and apparently in reasonable condition. This would generally include properties that:

- are of Victorian to present day construction;
- have load bearing structures or simple frames;
- use conventional building materials and construction methods;
- have service systems commonly used in domestic residential dwellings.'

There is further flexibility for surveyors who have particular technical skills and experience of 'unusual or specialist construction types', such as:

- 'properties (but not pre-Victorian) built using less common structures and materials. For example, 1920s timber framed 'mock Tudor' houses, thatched roofed [cottages];
- prefabricated concreted and steel framed properties built after the First and Second World Wars;
- properties using new and developing technologies or materials.' [For example, modern methods of construction, the one-off 'grand design' dwellings, and so on.]

Although these explanations are useful, that still glosses over a number of important issues:

1. *'Particular technical skills and experience'* – According to the practice notes, this has to be a proven specialism, but in my view it does not have to be something you do daily. It could be something where, if challenged, you are able to justify your decision to take on the task.

Justification could typically include:

- a life long learning and professional development strategy which reflects that type of work;
- a portfolio of previous surveys and other types of instructions showing your experience.

I always ask myself one question: 'Would I be happy defending this in a courtroom?' If the answer is 'no', then you should turn down the instruction, no matter how welcome the work would have been.

2. *'Pre-Victorian'* – At the moment no dates are specified (although we know that Victoria was crowned in 1837). As this book goes to press, RICS has indicated that these age descriptions may be broadened. Therefore, if you are unsure, always check the most recent version of the practice notes.

3. *'Prefabricated properties'* – There are hundreds of different types of prefabricated property in the residential sector. Some are widely recognised and have good supporting technical advice from government agencies (for example Cornish, British Iron and Steel Federation, and Unity dwellings), while others are so restricted in distribution and number that few surveyors even know of their existence, let alone their condition.

Ideally, for each type you claim to know, you should:

- be able to recognise and identify that particular type, together with all its variants;
- know whether it was designated under the *Housing Defects Act* 1984;
- be able to identify the problems associated with that system;
- be familiar with the usual range of approved repair systems and their general acceptance for mortgage purposes; and
- be able to advise on the implications and risks in connection with the mutual support between adjoining or adjacent properties.

You are not expected to carry all this information around in your head; even the most knowledgeable among us will have to check a few facts and details. But if you are answering 'no' more than 'yes' to these questions, then you should consider whether to accept the instruction.

4. *Historic and other traditionally constructed buildings, listing and conservation areas* – I mention this because surveyors often see a building survey as the only option for listed buildings. Where the property is a 17th Century cruck-framed cottage I would agree. However, 'listing' protects a broader range of buildings including the more familiar 'traditional' buildings. If you are asked to do an HBR on a listed building then you should ask whether you are familiar with:

- the likely method of construction and materials, especially those that affect the building's ability to 'breathe';
- typical modern repairs that can damage older buildings (for example, cement based mortars and renders, injected chemical damp proof courses and replacement concrete floors);
- the rules and regulations governing the use and development of listed buildings and those in conservation areas; and
- advice and guidance about the implications of owning a listed building.

Even if you do have specialist knowledge and expertise, you must always make sure you stay within parameters of the new HomeBuyer Service. For example, you may be able to take concrete samples from a prefabricated home or expose the concealed sole plate of an older timber framed building but this must be capable of inclusion under 'Additional advice' in the HBR and agreed in writing with the client.

Finally, court precedent and good practice requires surveyors to be familiar with the types of buildings that are common in the geographical area where they work. Traditionally, the consensus was a surveyor should practice within a 25-mile radius of their office. Although the practice notes do not state a specific distance, paragraph 1.4 [page 4] does state that such knowledge will typically include:

- 'common vernacular housing styles, materials and construction techniques;
- environmental issues, including flooding, aircraft noise, radon levels, mining, soil conditions, major areas of potential contamination, etc;
- the approximate location of the main conservation areas/historic centres;
- local and regional government organisations and structures;
- an awareness of the socio/historical/industrial development in the area;
- the housing market and property values in the area.'

The practice notes also state: 'If the surveyor does not have all this knowledge, they must fill the gaps through thorough research. If this cannot be achieved within an appropriate timescale, the instruction should not be accepted.' [paragraph 1.4, page 4].

PRE-INSPECTION CHECKS

Knowing the geographical area is crucial when it comes to pre-inspection checks.

Section D of the HBR asks for a variety of general information about the property, including spaces for details of 'outside the property', 'the location', 'facilities' and the 'local environment'. The practice notes [About the property, page 42] state that any comments in the HBR should be based on the surveyor's inspection and local knowledge. Furthermore, the 'Description of the RICS HomeBuyer Service' [HBR, page 24] says: 'The surveyor will not make any enquiries about contamination or other environmental dangers. However, if they suspect a problem, they should recommend further investigation.'

This clearly establishes that surveyors do not have to carry out a property-specific search for every job they do. However, I think it is right to retain some flexibility. Depending on experience, some surveyors will have extensive local knowledge built up over years of practice, while others – either early in their career or new to an area – may need to carry out property-specific research to plug their knowledge gaps. This need not be an extensive desktop study. There is so much useful, free information available that can help add depth and breadth to your inspection and report.

The other reason for a little background research before you go is that it adds to your inspection 'agenda' and helps to make sure you collect the right information. For example, you might carry out an inspection and return to your office only to find that there is an urban stream close to the property which is likely to flood the area. Better to have known this before you went so you could check the local topography. In some cases, it can help you avoid a claim.

However, there is a danger that if you exceed the role described in the practice notes you run the risk of raising the expectations of your clients and possibly leave yourself open to legal challenge. It is the familiar argument of: 'If they told me about that, why didn't they tell me about this as well?' However, if you miss something that a two-minute Internet search in the public domain could have revealed, the courts will show little sympathy. In an effort to help you through this minefield, I have identified what I think are the most important sources of information that can help you develop your 'local knowledge'.

Environment Agency

The Environment Agency (EA) (www.environment-agency.gov.uk) is the public body that protects and improves the environment and monitors the air, land, and water in England and Wales. Its website includes information on: flood risk; pollution risks; landfill; and ground water protection zones.

Warning: Information from the EA's site does not constitute an 'environmental assessment' or make you an expert. Instead, it will give you a partial insight into the nature of the locality. It may help you spot features and issues that would ordinarily have remained hidden or unknown.

To search any of these categories, you will need the correct postcode of the property. The best place to begin your search is on the 'home and leisure' page under the heading 'What's in my backyard?'. Here, among other information, you will find:

- **Flooding maps** – These maps will help you find the predicted risk of flooding in the area, together with the instructions on how to use them.
- **Pollution** – The main source of pollution is sewage, but other common pollutants include asbestos, household rubbish and vehicle parts. This map shows the EA's latest pollution data including air, land and water incidents.
- **Landfill** – Here, there are two main types of information:
  - **Active landfill sites** – These are sites where local authorities and industry can take waste to be buried and compacted with other wastes. The EA issues licenses and regulates these sites to ensure that their impact on the environment is minimised. The maps show the boundaries of each landfill site drawn from the plans. However, they admit the small scale of the maps makes it hard to be 100 per cent accurate. The boundaries can follow field boundaries or roads, but in most cases are within five metres of the actual boundary.
  - **Historic landfill sites** – These are locations where the site is now closed and/or the waste is covered. This is based on information from data held by local authorities, the former Department of the Environment, British Geological Society and from EA suspended authorised landfill licences.
- **Groundwater protection zones** – The EA has defined Source Protection Zones (SPZs) for up to 2000 groundwater sources such as wells, boreholes, and springs used for public drinking water supply. These show the risk of contamination from any activities that might cause pollution in the area. The closer the activity is, the greater the risk. The maps show three main zones (inner, outer and total catchment). These are used to set up pollution prevention measures in areas that are at a higher risk, and to monitor the activities of potential polluters nearby. This map will be useful for surveyors working in rural areas, where ground source water supplies are more common.

Defra noise maps

In July 2008, the Department for Environment, Food and Rural Affairs (Defra) published noise maps for the major cities, rail routes, and airports in England. These can be downloaded free of charge from www.defra.gov.uk. The strategic maps are to satisfy the requirements of the Environmental Noise (England) Regulations 2006, and estimate noise levels from:

- **major roads** – roads with more than six million vehicle passages annually;
- **major railways** – railways with more than 60,000 train passages annually;
- **major airports** – airports with more than 50,000 aircraft movements annually (except for training on light aircraft); and
- **urban areas** – with populations greater than 250,000 and a certain population density (including, for example, Blackpool, Bournemouth, Brighton, Bristol, Coventry, Hull, Leicester, Liverpool, London, Manchester, Nottingham, Portsmouth, Preston, Reading, Sheffield, Southampton, Southend, Teesside, The Potteries, Tyneside, West Midlands).

The noise maps are like a 'weather map' for noise: they show areas that are relatively louder or quieter. The maps have two main purposes: to provide information on noise levels that can be linked to population data, and to help in the production of noise action plans so the authorities can manage the problem and reduce noise levels where appropriate.

Although the maps have not been specifically designed for surveyors, a quick postcode search can reveal the noise 'bandings' around major routes that are not always obvious during the site inspections. The noise contours around airports and rail routes are particularly useful.

Defra points out that the maps are only intended to be used for strategic assessment of noise levels in any given area and should not be used to attempt to '… determine, represent or imply precisely the noise levels at individual locations (e.g. individual houses, windows).' (Source: http://noisemapping.defra.gov.uk/wps/portal/noise)

The maps show average noise levels for an average day in the year, and do not show the specific noise from individual vehicles, trains, or aircraft or from discrete industrial activities. Therefore, the information must be used with great caution. However, it can give a relative measure of how noisy a road or railway could be – a factor that is difficult to assess if you are conducting the survey outside the rush hour or during school holidays.

Geological information

Although the practice notes do not include a requirement to assess the likelihood of subsidence of the property, a general knowledge of your local area is important to understand the geology of the sub-soil.

There are two useful sources of information:

The Coal Authority

This is a government-funded organisation that aims to '… facilitate the proper exploitation of the Nation's coal resources, whilst providing information and addressing liabilities for which the Authority is responsible, in a professional, efficient, and open manner.' (Source: www.coal.gov.uk)

The Coal Authority holds and maintains the national coal mining records as well as the Cheshire Brine Subsidence information. Since October 2006, this information has been available for specific properties. Although surveyors are not specialists in this area, we should at least know whether there has been mining or brine pumping in the area we cover. The Coal Authority's website (www.coal.gov.uk) can give you an indication of whether areas need a mining report by three different methods (see box).

Mining reports

At the time of writing, the Coal Authority website listed the following 'affected areas' for English and Welsh counties (tin mining is not covered by this list).

Bristol	Lincolnshire
Cheshire	Merseyside
Cumbria	Northumberland
Derbyshire	Nottinghamshire
Devon	Shropshire
Durham	Somerset
East Riding of Yorkshire	Staffordshire
Gloucestershire	Tyne and Wear
Greater Manchester	Warwickshire
Herefordshire	West Midlands
Kent	Worcestershire
Lancashire	Yorkshire (north, south and west)
Leicestershire	

Apart from the two Welsh counties mentioned below, all other areas in Wales are affected.

The following areas do NOT need a mining report:

Bedfordshire	Hertfordshire
Berkshire	Norfolk
Buckinghamshire	Northamptonshire
Cambridgeshire	Oxfordshire
Cornwall	Rutland
Dorset	Suffolk
East Sussex	Surrey
Essex	West Sussex
Greater London	Wiltshire
Hampshire and the Isle of Wight	Ceredigion (Wales)
Gwynedd (but not Isle of Anglesey)	

Alternatively, the Coal Authority provides:

- The Gazetteer – an 'A-Z' of place names, with mining reports being recommended for those areas that are highlighted in bold on the list; and
- Property Search Service – a postcode search facility that will simply reveal whether the Coal Authority recommends a further mining report, but which provides no other information.

The Coal Authority charges £20–£45 for these products, and it is not surprising further reports are often recommended.

British Geological Society

A more detailed picture of the prevailing ground conditions is produced by the British Geological Society (www.bgs.ac.uk). The BGS website includes a whole range of postcode specific products including geological maps for the area, natural ground stability, and radon reports, through to a detailed assessment of whether the location is suitable for a ground source heat pump! These are very expensive products and, like the Coal Authority reports, outside the remit of the HBR service. However, if you are getting to know a new area or building up your knowledge after recently qualifying, you could consider buying a few of these reports for your own purposes.

Health Protection Agency

The Health Protection Agency (HPA) (www.hpa.org.uk) is an independent body that aims to protect the health and well-being of the population. The HPA plays a critical role in protecting people from infectious diseases and preventing harm from hazards involving chemicals, poisons, or radiation. The HPA publishes information on a number of topical issues on its website:

Radon

The Indicative Atlas of Radon in England and Wales (JCH Miles et al., November 2007), published by the Health Protection Agency and the British Geological Society, is based on a detailed map assembled from data obtained from radon tests of over 460,000 homes. The information is presented on the Atlas as one kilometre squares, coloured according to the highest radon potential it contains. It is important to note that, if a specific property is within an area on the map that has a higher percentage of homes above the 'action level', this does not necessarily mean the particular house is similarly affected. It could be many hundreds of metres away from a small area of high exposure. Therefore, you should use the Atlas as a trigger for a more accurate search.

A more precise picture can be obtained from www.UKradon.org, a site run by the British Geological Society and the Health Protection Agency, which provides radon risk reports and measuring services. Here you can obtain an estimate of the probability that an individual property in England and Wales is above the 'action level' for radon. However, there is a charge for this service, although at the time of writing, this is less than £5.00.

The Building Research Establishment (BRE) also offers useful advice for homeowners at www.bre.co.uk/radon. The advice is based on the type of searches offered at the www.UKradon.org site. Surveyors can use this advice to create very robust paragraphs for their own reports.

OFCOM (mobile phone base stations)

The Sitefinder website (www.sitefinder.radio.gov.uk) is a national database of mobile phone base stations and their emissions, administered by OFCOM. This database was set up following investigations into the possible hazards posed by mobile phone technology such as base stations. The health effects of mobile phone masts are still very much open to debate, so you should keep developments in this area under close review. I do not want to contribute to this debate, but many homebuyers will simply want to know if a mast is just around the corner.

Free environmental risk assessment services

Several websites allow users to review features and characteristics of their own neighbourhoods. The best known of these is www.homecheck.co.uk, a free database that can be searched by postcode. This can give you an indication of whether the neighbourhood is at risk of flooding, landslip, radon gas, built on a landfill site, suffers from subsidence, or close to an area of industrial pollution. School performance ratings, crime rates and local amenities, as well as a summary of large and small planning applications in the area. can also be obtained.

Use the data carefully. The site operator acknowledges that it offers a general view of environmental risks and planning information in a particular area, but none of it is property-specific. It is '... not intended to be a replacement for a professional environmental and planning report' and should not be relied upon as a precise assessment. Note also that the website can only be used once a day, and is not always accessible. You can buy accurate assessments from the parent company at www.land markinfo.co.uk

In my view, you should not rely solely on free resources while assembling information for a survey report. Here is a cautionary tale:

I carried out a search of a property that I often use in my training seminars. This revealed 'high risk' warnings for: flooding; coal mining; subsidence; past landfill sites; historic land use (past industrial use); and air pollution. But from my local knowledge I know the area is not the toxic wasteland that these results suggest. The property is not at risk of flooding because it is approximately 25m above the extreme edge of the identified flooding zone. Although it is in a former coal mining area, the last extraction was over a century ago and would present no risk at all.

I have no knowledge of subsidence problems in the area and I was unaware of any problems with the subsoil. Historic maps reveal that the area has been heavily developed but this did not reveal any obvious historic landfill sites nearby. The property is close to the centre of a former industrial city and inevitably will be near previous industrial uses, but these were likely to be smaller scale operations and so unlikely to present a serious risk. However, the city centre location of the property does result in high levels of air pollution. From the six aspects identified by the free websites as 'high risk', only one would need further consideration. This shows the importance of local knowledge.

However, if your client uses the same sites and sees six 'high risk' warnings, they will wonder why your HBR mentions nothing about them. Be ready with your explanation.

Regional construction techniques and local knowledge

In some areas of the country, building practices have evolved in an insular way so that they are geographically restricted to small areas of the country. For example:

- In many western and northern areas of the country, cavity walls have been used since the early 1900s. In parts of the south east, they were not common until after the Second World War.
- In a small part of the West Midlands, builders regularly used solid brick walls (usually Flemish or English bond) for the front elevation of detached houses and cavity construction for the other walls. In other parts of the country, I have seen the complete reverse!
- In the Brighton area, 'bungeroosh' walls are made out of random (and often poor quality) mixes of lime mortar, flint and just about any other material that was laying around.
- The use of 'cob' (i.e. mud and straw) for wall construction is very common in Devon.

Being familiar with local construction techniques can help you identify property quickly, effectively and with confidence, and it is the first step to assessing the condition of a building. Here are a few examples of where to look to gain this degree of familiarity:

- **Local and regional publications** – Typically, these describe building techniques for that area or the history of local urban development. Because of restricted circulation, these may be difficult to track down but you can often find them at National Trust bookshops, local museums, craft centres, local history societies, and so on.
- **The Society for the Protection of Ancient Buildings (SPAB)** – This national organisation has a regionally based membership. Members are very knowledgeable about local construction characteristics and are usually happy to share that knowledge at meetings and lectures. Many of the meetings include practical demonstrations of how to use older building materials. See www.spab.org.uk
- **Internet sources** – Identifying reliable sources of information on the Internet always presents a problem, because they can change regularly and the quality of their content varies tremendously. Some website owners have researched the information well, while others are at best anecdotal and subjective. However, websites such as 'Bricks and Brass' (www.bricksandbrass.co.uk/index.html) and 'Researching Historic Buildings in the British Isles' (www.buildinghistory.org) are useful and provide many links to other resources.

Listed buildings and conservation areas

It is important to identify whether the property is listed, or in a conservation area. Some very modern buildings have been listed because of design or technological innovation, or because of an association with well-known characters or events.

The following guidelines can help you to identify a building that is protected by listing:

- All buildings built before 1700 are likely to be listed.
- Most buildings built between 1700 and 1840 are more than likely to be listed.
- Between 1840 and 1914, only buildings that are of definite quality and character are listed (but bear in mind that they may be sited in a conservation area).
- Properties built after 1914 usually have to be outstanding before they are listed.
- Buildings that are less than 30 years old are only listed if they are regarded as having significant architectural merit and generally under threat.

You can find information on listed buildings from www.imagesofengland.org.uk, where you will need to register to gain a password for future free searches. Alternatively, you can contact the local planning authority, which will have a record of all the listed buildings and conservation areas within their area. If you are fortunate, they may have this information on their website so you can search the database remotely.

By obtaining general information like this you can build up a body of knowledge of your own area that can reduce the chances of you being caught out.

TOOLS AND EQUIPMENT

Unlike the guidance within some of its predecessors, the HBR practice notes do not provide a detailed list of the equipment a surveyor should take on a survey. Instead, the practice notes state:

'The list of equipment is not intended to be comprehensive. At their discretion, surveyors may – but are not obliged to – use other equipment (e.g. compass, tape, inspection cover lifter, meter box key, spirit level). The use of cameras, while encouraged, is discretionary. There is no objection to the use of machines for recording site notes, provided that a permanent record of those notes is kept.' [practice notes, page 27].

The word 'discretion' gives you a degree of choice, but you will need to take equipment that enables you to complete the inspection correctly. Using the 'Description of the RICS HomeBuyer Service' (referred to as the 'DHS') [practice notes pp24–27] as a guide, I think you should have the following equipment available to you on every inspection.

Binoculars

These can either be standard size or compact binoculars (the ones that fit in your pocket), as long as they are 10× magnification. These are essential so you can see the roof, chimney, and other upper building elements from a distance.

Ladder

The DHS states the surveyor '… may use a ladder for flat roofs and hatches no more than three metres above level ground (outside) or floor surfaces (inside) if it is safe to do so'. In my view, this sentence is written for public consumption, because a surveyor may, and occasionally may not, use a ladder during the inspection. For example, where there is no need to use a ladder (in a flat) or it is not safe (see the discussion below) and the client becomes aware, they could mistakenly assume you have not done a proper job. However, from a surveyor's point of view, you should always use the ladder where appropriate and it is safe to do so. You should not see the DHS sentence as a suggestion that the use of the ladder is optional!

To use the ladder safely, the top portion of the ladder should extend 1.0m above the stepping off or viewing level. Consequently, the ladder should be at least 4.0m long, and I suspect those owned by most experienced surveyors will be less than this. This presents those of us with a shorter ladder with a dilemma: if we do not inspect a 3.0m high loft hatch for health and safety reasons, the courts will take a dim view. However, if we do use a ladder that does not extend into the loft sufficiently and fall off, the insurance company may refuse to help. This may be a good time to consider buying a new ladder.

Measuring devices

I think a 5–7.5m retractable steel tape is suitable for most inspections. Some rooms will often have one dimension that is greater than five metres, making the additional length useful.

Personally, I have recently discovered the delights of handheld laser 'distance meters' and discovered (a fact everyone knew) they really do speed up inspections. However, there will be occasions when there is no opposing wall to 'bounce' the laser off, so you will always need a measuring tape as a back up. Externally, you may need to take longer measurements to check the relative positions of boundaries and so on, for which a 20–30m fabric tape will be required.

Where the property is affected by structural movement, a short steel rule marked in millimetres (the type used by engineers) can help you record these more accurately. I always take a 2m folding 'boxwood' rule and find it excellent for measuring heights of rooms and other features that are sometimes difficult to measure with a floppy steel tape.

Electronic moisture meter

I always recommend using a conductivity moisture meter (the one with steel pins), because these are generally more reliable than the 'flat plate' varieties. You should always carry spare batteries and calibrate the meter in accordance to manufacturer instructions every time you use it.

Torch

A powerful 'mini' flashlight is acceptable, but I prefer a professional-sized torch (the type that takes two or three D-type batteries). These are more robust and give a reliable light source. You should always carry spare batteries and bulb.

Lifting equipment/crow bar

The main purpose of this equipment is for lifting inspection chamber covers. A 'crow bar' (minimum length of 450mm) is sensible because chamber-cover lifting 'keys' rarely work as the lifting holes are always rusted or full of debris. To make the lifting of covers easier, I think that a few other tools can help:

- robust claw hammer or small (2kg) club hammer;
- large flat-head screw driver, for levering up the cover; and
- bolster or cold chisel, for loosening up the most difficult of covers.

These additional tools are also useful for other simple opening up tasks, such as lifting loose floorboards.

IMPORTANT NOTE: By identifying these tools, I am not suggesting that the new HBR goes further than the old HSV and involves 'opening up' work; merely that these tools can make lifting a loose inspection chamber lid easier and safer.

Spirit level

I think two spirit levels are useful:

- A small hand-held level (sometimes called a 'boat level') is useful for checking alignment of doorframes, windowsills, etc.
- A 1–1.5m 'bricklayer's level' will be essential to check the levels on floors and the verticality of walls – the longer the better (but makes sure it fits in your car).

Many surveyors carry a marble or golf ball to assess whether the floors are sloping. The drawback is that it does not work very well on deep pile carpets. Even if this does reveal a slope, it is difficult to assess its seriousness. The best method is to use a longer spirit level, place it on the floor and if you insert 'shims' (for example, pieces of thick cardboard) under one end until it is level you can then measure the gap to the floor surface at the other end. A simple calculation will give the precise slope.

Pocket mirror

This is a very useful instrument that can help you look under floors, within ducts and inside other voids. I have a 'search' mirror that I can fix on a telescopic arm and this extends my field of view.

Plumb bob

This can provide a quick and effective way of checking the verticality of walls, especially when dropped out of a convenient window. A 1m spirit level gives some indication of verticality, but a plumb line is much better. When on your own, you will not be able to take precise measurements between the string and the wall, so this method can only give you a visual indication of verticality.

Compass

This is an essential tool. It can indicate the direction of the prevailing weather – important information when looking for dampness.

Digital camera

Some report-writing software programs will allow you can add photographs at the end of the report, but the images are not a formal part of the report. The practice notes state that '… the use of cameras, while encouraged, is discretionary' [paragraph 7, page 27]. However, most surveyors have embraced the digital revolution and use cameras that are able to take and store a large number of images. This is very useful when you are writing or reviewing your report, but it can also be crucial when responding to an enquiry or a challenge years later. Nevertheless, digital images are no substitute for adequate and clear site notes. If you do take photographs, it is important to store them securely where they can be easily retrieved.

Protective clothing

To many surveyors, my advice to take protective clothing – such as overalls, umbrella and wellingtons – will be self-evident, obvious, and possibly mildly offensive. My reason for including this relates to the very nature of the HBR itself. My years of training residential surveyors have led me to a few anecdotal observations:

- Most 'level two' inspections are carried out by individuals who usually carry out mortgage valuations or their equivalent. Because these 'level one' products are limited in their scope, most practitioners will carry out the inspections in their working clothes (suits, smart jacket and trousers, etc.) reasoning that they are unlikely to get too dirty.
- When they switch to the occasional 'level two' product, many do not change their routine – or their mindset. For example, a 'head and shoulders' can be done in your office clothes, but where you have to get off the ladder for an HBR (as long as it is safe to do so) to inspect the roof space more thoroughly, overalls would be a sensible choice.
- After an hour and a half of focused concentration in the house, surveyors can be too casual when carrying out the external inspection, especially when it is raining and muddy underfoot. Outbuildings, boundaries, retaining walls, external garden lighting, and ponds can have an impact on both the safety and value of the property. Being able to inspect this comfortably in poor weather is an important part of the service.

I am not suggesting that the HBR inspection has changed in any way but that, in my opinion, many practitioners approach the inspection as if it is a slightly longer version of a mortgage valuation. This is not the case. During an HBR (as with the HSV), you may have to change your jacket for your overalls and stay out in the rain longer than you would normally choose. You should be prepared to do this by ensuring that you always have appropriate clothing with you.

Health and safety equipment

On an inspection, you will need items to protect your own health and safety. As experienced practitioners, you should be aware of these, but I have provided a comprehensive list for your information:

- A personal attack alarm (just in case of emergencies) and spare battery.
- A mobile phone so that you can call for help and advice if there is a problem.
- A first aid kit that suits your particular professional role. The basic minimum suggested by the Health and Safety Executive (see the HSE leaflet INDG214) includes:
  - a leaflet giving advice on first aid;
  - 20 individually wrapped sterile adhesive dressings;
  - two sterile eye pads;
  - four individually wrapped triangular bandages;
  - six safety pins;
  - six medium-sized individually wrapped unmedicated wound dressings;
  - two large unmedicated wound dressings;
  - one pair of disposable gloves.
- A safety helmet – not required for most normal inspections, but where you are looking at a partially complete, renovated property or one in poor condition, a helmet should be worn.
- A facemask with disposable filters for loft inspections and other dusty environments. It should be suitable for general use and does not need 'asbestos' grade filters because you should not go into spaces that contain asbestos products. If you think there is a strong possibility of asbestos dust then you need to leave the area immediately (see 'Health and safety matters', below).
- Protective gloves for lifting inspection chamber covers, etc. Some organisations are now recommending disposable latex gloves because if reusable gloves are used to lift an inspection chamber cover and simply thrown back into your tool bag, they could contaminate other equipment.
- Safety shoes that have appropriate toe and sole protection for use when inspecting partially renovated properties or those in a poor condition. Wellington boots (with similar sole and toe protection) are good for very muddy gardens and grounds.
- Disinfectant hand wipes to clean your hands after an inspection (and especially before you eat your sandwiches).

HEALTH AND SAFETY MATTERS

Over the past 30 years, health and safety legislation has become increasingly central to our everyday activities. 'Risk assessments' have become so widespread that they are often the target of comedians on satirical radio and television shows who see health and safety procedures as examples of the 'nanny' state intruding on our daily lives. However, accidents at work are no joke. Although surveyors do not use dangerous machinery or go down deep holes, we do use ladders every day and regularly visit building sites. We often meet aggressive dogs, not to mention their occasionally belligerent owners.

Fragmentation of our sector is another – often overlooked – threat to health and safety. As larger surveying organisations have downsized, more surveyors have become self-employed 'consultants'. Although this provides a more responsive and flexible workforce, it also weakens sector-wide policies and procedures such as pension provision, health and safety, quality assurance, and so on. Self-employed surveyors worried about work continuity can regularly accept too many instructions in an attempt to build up reserves to help them get through leaner times ahead. In this situation quality might suffer, but more importantly, hard working, tired surveyors will often take silly risks.

This section reviews the legal framework and some of the most important issues that affect surveyors in their daily work. This is important for newly qualified surveyors, but it also applies to those who have been carrying out the job for years. My experience suggests that older practitioners developed their inspection methodologies many years ago and although sound, they might not have accounted for current philosophies and regulations.

The legislation

The *Health and Safety at Work Act* 1974 is the most important single piece of legislation and sets out the general duties of both employers and employees and the obligations they owe to the public. The Health and Safety Executive (HSE) is the government agency that administers this legislation, and their website includes additional information on the Act and other types of guidance (www.hse.gov.uk).

This is a complex piece of legislation, supplemented by a plethora of secondary Acts, regulations, and codes of practice. It requires that you identify risks to health and safety and take measures to minimise those risks.

The Act invokes a principle of 'so far as reasonably practicable', which means that not all risks have to be eliminated. Instead, it requires measures that are 'proportionate' to the risk faced (i.e. sensible and reasonable including cost).

For example, if an accident is very likely to occur then an employer/employee must take action to prevent it happening. A brick falling on an operative's head on a building site would be very likely and therefore, all operatives need to wear protective headgear. Conversely, encountering irritating dust particles during general domestic inspections, although possible, is not very likely. Therefore, surveyors do not have to wear a protective facemask all the time, but only if they go into areas that may be contaminated.

It is the application of 'common sense' rather than a prescriptive set of inflexible rules that underpins health and safety policies.

Although it may appear that the main legislation applies only to employers/employees, the self-employed should adopt a similar approach to safety – not least to ensure that they do not break the terms of their public liability insurance. Moreover, the self-employed have a clear duty to protect their customers and members of the public who they meet. They also have a 'moral' duty to protect themselves, especially if they have dependants who rely on their income. Remember, accidents can and do happen. For example, the HSE cites the case of a building maintenance worker who tripped on the second rung of his ladder. He caught his feet, fell backwards and hit his head on a concrete step, and died. The cause of an accident does not have to be dramatic; the simplest of mishaps can have serious consequences.

Employers and health and safety legislation

If you are an employer, then you must meet a number of clear responsibilities imposed by the legislation. For example, in order to meet the requirements of the Management of Health and Safety at Work Regulations 1999, an employer must:

- provide information about health and safety to their employees;
- have emergency procedures for accidents;
- carry out risk assessments to identify risks;
- eliminate and control risks ('so far as is reasonably practicable');
- maintain insurance for their employees and the public;
- carry out health surveillance of their employees;
- provide necessary equipment for the type of work involved;
- provide for those with special needs;
- control working hours and provide regular health and safety training.

If there are more than five employees, they must also have:

- a written health and safety document, e.g. a manual;
- a policy statement issued by a senior executive that is regularly reviewed and updated;
- health and safety roles and responsibilities within the organisation;
- management systems in place that plan and implement health and safety policies;
- a system that measures and audits health and safety performance.

It is beyond the scope of this book to offer further advice on the wide range of health and safety issues that are faced by employers of five people or more, and therefore you should contact the HSE directly with any additional concerns. However, there are some basic steps that everyone should follow (see box).

Risk assessment – the principles

The HSE publication *Five steps to risk assessment* (INDG 613 2006), sets out a risk assessment process that provides the main method of identifying hazards in the workplace. This assessment can be as simple as a careful examination of what could cause harm to people in the workplace. Once the assessment is complete, the employer should weigh up whether existing precautions are sufficient, or take further action, if necessary. The five steps are relevant to surveyors, and are therefore described briefly below.

Step 1 – Identifying the hazards

Identify how people might be harmed. One approach is to think through what you do as a surveyor. Although you need to include all potential hazards, stick to a 'common sense' approach and avoid risks that are too remote (unlikely). An important consideration for surveyors is the number of working environments they encounter:

- the office;
- travelling to and from the inspection; and
- the inspection itself.

All three must be assessed. Do not purely rely on your own assessment, as you may have missed something. The HSE suggests:

- ask other people working in similar environments, who might be able to identify hazards you have not thought of;
- visit the HSE website, where there are a number of publications relevant to the surveyor's working environment;

- seek guidance or advice from appropriate professional or trade bodies (for example, the RICS publication *Surveying Safely* is a comprehensive booklet that identifies the hazards surveyors may face);
- check any manufacturers' literature for descriptions of hazards associated with your equipment (for example, there is health and safety advice on ladders, mobile phones and even moisture meters in their respective user manuals).

Step 2 – Decide who might be harmed and how

For each of the hazards identified in Step 1, you should be clear who might be harmed (a simple note of the groups of people will suffice). For surveyors, this would typically include employees, fellow workers, occupiers of inspected property and their visitors, and members of the public. You should then identify how they might be harmed and what injuries or ill health could occur. For example:

- occupiers may be injured by items dropped by the surveyor while up a ladder;
- the surveyor could be injured by various pets/animals kept by the occupiers;
- the public could be injured by a surveyor's poorly maintained car if it crashes.

The HSE also points out that not all workers have the same requirements. For example, new and inexperienced surveyors might not be as competent at using sectional ladders as their more experienced colleagues and so pose more of a risk to themselves and others.

Step 3 – Evaluate the risks and decide on precautions

The next step is to decide what to do about the hazards. This is where the principle of 'reasonably practicable' comes into play. Although you can make these judgements yourself, it is most helpful to compare what you are doing against good practice and the sources of information identified above. Once you have done this, useful questions could include:

- Can I get rid of the hazard altogether?
- If not, how can I control the risk so that harm is unlikely?

By way of example, try applying these questions to the three hazards identified in Step 2:

Hazard 1. Occupiers may be injured by items dropped by the surveyor while up the ladder:

- You cannot get rid of the hazard altogether, loft spaces have to be inspected.
- Make sure you know how to use the ladder safely.
- Inform the occupiers that you are assembling the ladder and ask them to keep children/pets away from the area or under control.
- Do not take too many items onto the ladder with you. For example, do not take a clipboard, pen, torch and camera all at the same time. Not only does this increase the chance of dropping one but it also makes you less stable.

Hazard 2. The surveyor could be injured by various pets/animals kept by the occupiers:

- The risk cannot be completely avoided as many homes will contain animals of one form or another.
- When arranging the appointment, ask if there are any animals/pets and if yes, ask the owner to keep the pets under control and out of the way if possible.
- When arriving at the property, during the initial discussion ask the person in charge if there are any animals. If yes, ask for them to be out in another room or outside while you are inspecting.
- Make sure your tetanus injections are up to date, just in case of future mishap.

Hazard 3. The public could be injured by a surveyor's poorly maintained car if it crashes:

- Make sure the car is fit for the purpose, i.e. roadworthy.
- Make sure the car is properly and regularly serviced and that basic checks are carried out for every journey (e.g. windscreen washers, tyre pressures, etc.).
- Do not become distracted during the journey (for example, do not use mobile phones, eat or drink etc.)

In this way, you can take sensible precautions that do not necessarily cost a fortune but ensure the risk of anyone being harmed is kept to a minimum.

Step 4 – Record your findings and implement them

Once you have gone through this process, it is important to record the evidence. If you are self-employed you do not have to do this legally, but writing it down will help you review your procedures later. You might also be better able to deal with an insurance claim, or complaint, if you adopt this approach.

The HSE recommends these records should be short and simple and they have produced a pro forma for this purpose.

If you need to make improvements, the HSE points out that you do not have to implement them all at once. Tackling the cheap and easy things first, with temporary solutions for the most serious hazards, and with long-term plans to resolve the more significant problems, is a good first step.

Step 5 – Review risk assessments and update if necessary

In my experience, most practitioners miss out this stage, yet it can be one of the most important. Once you settle into a role or a function, your approach becomes almost second nature. However, circumstances and working practices change over time and if you do not keep these under review, new risks and hazards may not even be recognised.

The office environment

Although you will spend much of your time inspecting properties or travelling between them, you will also be office-based. Normal health and safety requirements will apply:

- adequate ventilation and air changes;
- comfortable working temperatures – for people who work in sedentary environments, the temperature should be at least 16°C;
- the provision of adequate light;
- facilities for waste disposal and maintaining cleanliness;
- the avoidance of cramped conditions;
- regulations about workstation design and seating;
- maintenance of equipment and services;
- safety of circulation space;
- the provision of toilet and washing facilities, drinking water, rest and changing facilities;
- maintenance of first aid materials and a file for recording all accidents that occur.

Much of your office-based time will be spent on the computer and the use of visual display units (VDUs) is particularly important. The HSE advises that you should use a well-designed workstation, with a chair that you can adjust to suit the table height and to encourage good posture. Laptops are particularly problematic, because manufacturers do not always design them to be ergonomic, so you should always use a desktop computer if one is available, or use a laptop with a separate screen and keyboard. You should also take regular breaks when using the computer.

A number of surveyors use 'tablet' computers or personal digital assistants (PDAs) to record site notes, uploading the finished report to a databank through a mobile phone from the front seat of a car. This can result in poor posture and back problems. For more information, the HSE's leaflet INDG36 is very useful.

Travel and general vehicle safety

Most surveyors spend a considerable amount of time in the car travelling between the office and inspections. You should ensure that the car is well maintained and roadworthy. In addition to the formal servicing, read the car maintenance manual and carry out routine checks.

It is also important to carry an emergency car kit in case of a breakdown, as well as a map to reduce the need of having to stop and ask for directions. A satellite navigation system is a good substitute and is now common in many vehicles.

You should plan your journeys so that you do not get tired. Take regular breaks on long journeys and follow the Highway Code. Accident statistics suggest that younger males are at greater risk during this part of the inspection, so kill your speed. Just a reminder: it is against the law to drive while under the influence of drugs or alcohol. It is also against the law to use a hand-held phone while driving. The Royal Society for the

Driving distractions

At the time of writing, the government had recently increased the penalty for drivers who cause death or injury while distracted. The definition of 'distracted' could include calling or texting on a mobile phone, drinking or eating, adjusting a satellite navigation system or car stereo, applying make-up or reading a map.

So how does this affect surveyors? To illustrate this, think about your own working day:

- You are likely to be frequently travelling between inspection appointments. You may be visiting properties in unfamiliar areas, so will have to refer to maps or satellite navigation appliances. The image of a surveyor driving slowly along a road looking at house numbers rather than the road in front is ubiquitous.
- If you do not have secretarial support you will have to deal with all enquiries and queries yourself. This could result in numerous phone calls throughout the day.
- Food and refreshments will usually be taken in the car between appointments.

The opportunities for distraction are considerable and in my view present a considerable risk. The Royal Society for the Prevention of Accidents (RoSPA) accepts that distraction is a difficult risk to manage but drivers can take some simple steps to avoid becoming distracted:

- If you need to do something distracting, find a safe place to pull over.
- Good planning can reduce the likelihood of becoming distracted in the vehicle.
- Recognise what distracts you – eating or changing a CD are examples of activities that drivers may do without thinking – self-assessment is an important part of developing your driving.
- Concentrate on your driving – easier said than done, especially in uninteresting environments.
- Make sure that you are ready to drive before setting off for a journey. If you are about to drive after an emotional event, then it is best to allow yourself time to cool down.
- Use technology sensibly. In-vehicle technology can be distracting, especially if there are several systems in the same car. For example, my satellite navigation system tells me which turning to take, bleeps when a speed camera is close and sounds an alarm when I accidentally (and rarely!) exceed the speed limit. In some urban environments, it can do this all at the same time – very distracting.

Prevention of Accidents (RoSPA) recommends that hands-free car phones should be avoided too, because of the risk of distracting the driver's attention from the road.

Home working

A growing proportion of surveyors now work from home, even when they are employed by a large organisation. In these circumstances, home workers are covered by health and safety legislation, and employers have a range of duties, just as they would if the staff were working in the head office:

- regular risk assessments of the home worker's circumstances – this should not be a generalised, standard procedure but one that takes account of the employee's particular circumstances;
- consideration of handling of heavy goods or equipment – surveyors will have to load and unload heavy sectional ladders and other equipment;
- provision of suitable equipment for the task in hand, and responsibility for its safe use and maintenance;
- periodic testing of electrical equipment – although the employer is not responsible for the condition of the electrical system in the employee's home even if they contribute towards the electricity bills;
- the safety of any hazardous materials provided for the work (for example, toner for printers and copiers);
- consideration of the hazards of computer screen use – this is likely to be one of the main issues for home working surveyors.

Risk assessments

Pre-inspection health and safety preparation

Before leaving your workplace, make sure somebody knows where you are going, what your plans are, what time you expect to finish the inspection and return to work or arrive home. And when the inspection is over, don't forget to check in with someone in the office to confirm you have finished, even if you then go straight home. A call from the safety of your armchair may stop your colleagues worrying about your whereabouts. If

...t work in an office, make a suitable ... with another associate, family member, ... ways carry a fully charged mobile phone and your personal attack alarm (but see my comments below on alarms). Go to the inspection in daylight wherever possible and organise your day so that you are travelling in darkness rather than inspecting buildings in a fading light. During winter months, this may mean completing the inspection by mid-afternoon.

On arrival at the property

You should undertake a risk assessment prior to starting an inspection. This does not need to be a formal process; you just need to take time to look around so you can make an initial assessment.

Appendix 1 includes a site notes template appropriate for surveyors. The second page of this includes a section titled 'Health and safety risks' where you can note down some of the more obvious potential risks. The risks should be recorded, the level of risk assessed (high/medium/low), along with how you plan to minimise the problems. This gives evidence that you have considered health and safety issues and may be useful if an accident does happen. Examples include:

- A roof in poor condition with a number of slates slipped and on the verge of falling off. This may be particularly dangerous during windy weather and may restrict your external inspections.
- An empty property in poor condition where dampness may have been penetrating for some time. This can result in timber defects to the floors, making the building dangerous to inspect.
- Large dogs or other threatening animals loose in the garden.

If you decide that the identified risks will limit your inspection in any way, it is important that you explain this in your site inspection notes.

Problems posed by the building

You must judge each property on its own merits, but the development of an inspection routine will help you recognise and minimise risks. Safety awareness has to be a top priority when entering any type of property, and experience will result in checks becoming automatic. If safety is not an essential consideration from the earliest stages, then this lack of awareness could lead to increased risk.

A comprehensive list of risks can be found in the RICS publication *Surveying Safely – Your guide to personal safety at work*, which is available free of charge from www.rics.org

Health and safety in occupied properties

Health and safety rules and regulations are usually associated with working environments where the dangers are more obvious. Inspecting residential properties is usually considered low risk. However, working alone in other people's property does have its own particular challenges that occasionally result in serious injury and even death. Organisations such as the Suzy Lamplugh Trust (www.suzylamplugh.org) and the HSE (www.hse.gov.uk) have produced excellent guidance for people whose work involves visiting people in their homes. Based on this information, the guidance notes issued by a number of employers, and my own experience, I have outlined below what I feel is a framework for a safe working environment for surveyors.

I do not want to fuel the anxieties of newly qualified surveyors. The fear of crime is usually out of all proportion to the likelihood of the events actually happening. However, unless you are aware of the dangers and plan accordingly, you could be vulnerable. The potential for violence may depend on a whole range of factors that could be nothing to do with you. The reason the property is on the market might be a result of a marriage break-up, bankruptcy or bereavement; the seller might be very worried about the effect that a negative survey report might have on the value. Therefore, you must always remain alert to the possible dangers. The Suzy Lamplugh Trust advises that you should do what is necessary to protect yourself, trust your instincts, and act accordingly. If you follow these suggestions, they will become a part of your normal inspection process and help you stay safe.

Before arriving at the property, think about its location (whether it is in a high-rise block, on a country lane or on a one-way street) and if there are any associated dangers. Consider whether you have to go alone. This may be difficult when resources are scarce and workloads are heavy, but two people will be much safer than one. In many cases, accompanied inspections can be part of a new entrant's training programme or justified by formal quality assurance requirements. If this is difficult to arrange, two person inspections could be restricted to areas where the risk is the greatest.

When you arrive at the property, park in a position where it will be easy to leave quickly. Do not park in a tight spot where it might be difficult to get out of in a hurry. Always lock your car and put any valuables out of sight in the boot.

When you are in someone's property, it is important to remember that it is their home and you are imposing upon their territory.

When gaining access to the property, state who you are, why you are there and show the person your ID or business card. Check who you are talking to and make sure it is the same person with whom you arranged access. If not, consider carefully whether you should go

Dealing with dangerous dogs

If dogs or other animals appear threatening, politely ask that the owner keep them out of the way. It is surprising how many dogs that 'wouldn't hurt a fly' have bitten surveyors. A participant at one of my training courses (who is a gas engineer) told a chilling tale of when he was attacked by the occupant's large dog. It sank its teeth into his thigh, and forced him to the floor. The only way he could get the dog to stop was by repeatedly stabbing the brute with a large screwdriver he was using to dismantle a gas fire. The engineer required 17 stitches to his leg and the dog suffered serious injury.

It is also important to ask whether there are animals outside before you inspect the garden/grounds. If you do meet an aggressive dog or other pet, try to stay calm and assertive. The best advice I've been given on what to do if a dog menaces you is:

- do not move suddenly;
- put hands in pockets (waving arms become a target);
- avoid eye contact, as that will be confrontational – look slightly away from the animal;
- back away but do not run – never turn your back to the dog, it is best to turn sideways;
- children can accidentally provoke a dog so be careful if there are both in the property;
- try not to allow an attack on your legs; if you go down the next bite could be your face;
- if the dog is large and heavy, brace yourself so that its pounce does not knock you over;
- if the dog forces you down, roll over onto your stomach and cover your head with your arms.

Although these suggestions may appear macabre, they help dispel common myths. How many times have I heard surveyors say 'You have to stand your ground and show no fear!' This is easy to say but hard to do!

A number of manufacturers sell ultrasonic dog deterrents that many lone workers carry with them. The high frequency sound bursts are inaudible to humans but apparently clearly heard and disliked by dogs. As a result, the manufacturers claim the animal stops, makes a detour, or retreats. I have no experience with these instruments but I know a number of people who claim they are effective. Whatever your view, you should always take the general precautions described above and if an ultrasonic device enhances your safety then that is a bonus.

---

in. In some cases, the property may be occupied by a tenant who may know very little about the building or even the fact that it is about to go on sale. In these circumstances, an accompanied inspection with the property owner will always be better.

If the only person in the property during the inspection is below the age of 16, or is someone who you judge to be vulnerable in some way, you should postpone your inspection until an adult can be present.

When carrying out the inspection, do not let the occupier lock the doors behind you. A night latch can be easily opened, but a mortise lock that you can only open with a key will restrict your ability to leave the property. Try to familiarise yourself with the layout of the property so you can get out quickly if necessary, and do not leave possessions in several different rooms as this may delay your departure if you need to leave in a hurry.

Do not take documents into the property that you do not want the occupant to see, and be careful when using a personal recorder to dictate notes. You may offend the occupiers by even the most objective assessments of their home. Different people live in different ways and try not to react to the condition of the property, even if it is very untidy, smelly, or just plain dirty. Do not judge and if you do, try not to let your feelings show.

Remember to keep your car keys, equipment, mobile phone, and/or valuables safe during the inspection. It is best to keep them with you. Many surveyors carry them around the property in a large shoulder bag so they are always on hand and can be packed away quickly if you need to leave promptly.

Make every effort not to damage the property. I know this is stating the obvious, but it is inevitable that you will occasionally enter a property with dirty shoes or knock an ornament off a shelf with your clipboard. Here are a few tips:

- If it is wet outside, or you have dirty shoes, either use non-slip plastic overshoes or take along a pair of lightweight shoes such as trainers or plimsolls to change into.
- If there are ornaments stopping you using your measuring tape, ask the occupier to move them or, if it is more appropriate, ask the occupier for permission, then gently move the items out of the way yourself and replace them when finished.
- Do not 'force' things to open: you might not be able to shut them again. Typical examples include meter and store cupboard doors, boiler flaps and windows. Ask the occupier to do it instead.

If you do damage anything, it is essential that you inform the owner or occupier immediately, because you will have public liability insurance that will cover major incidents. Where the damage is less serious, you need to settle the matter quickly and effectively to make sure you maintain a good relationship with the seller.

Remember that manners cost nothing, but can pay dividends. Before you use washing or toilet facilities, make sure you ask the occupier first. Do not probe into the occupant's private matters or belongings. Several years ago, an investigative TV documentary installed hidden cameras in an empty, occupied house and filmed a number of surveyors carrying out a survey. This revealed one surveyor who searched through the occupant's entire video collection and another who read all the messages and other information on a small notice board in the kitchen. This is intrusive and unprofessional.

If you accidentally enter bedrooms, bathrooms or WCs where occupants are sleeping or changing, apologise and leave the room quickly. On one of our training courses, a surveyor told us about his experience with a naked man in a cupboard. Apparently, the man was in his bedroom (presumably dressing) when he heard the surveyor coming along the landing. He was so embarrassed that he hid inside the wardrobe. The surveyor entered the room, started his inspection but on hearing a noise from the wardrobe he opened the door. Although this sounds amusing now, imagine how differently this could have turned out. Surveyors need to be ready for anything.

If you notice valuables around the property, you should consider asking the seller to remove them. This highlights the importance of walking around the property with the occupier so these problems can be spotted and resolved early on. If the property is unoccupied, you should make a record in your site notes of the time and circumstances when you first noticed the valuables and I would also advise you to phone your office or a colleague so that a third party has at least some evidence of the incident. Taking a photograph of the items will also be beneficial.

If any of the occupants appears to be drunk, drugged, aggressive, or overly attentive, then give your apologies and leave immediately. Most commentators say that if you feel uncomfortable about a person's attitude or behaviour, make an excuse, and leave straightaway. Do not worry what people may think of you; it is important to maintain your own safety. A number of organisations provide their employees with personal attack alarms and although these can be effective, the Suzy Lamplugh Trust takes a more balanced view:

- In an emergency, will you be able to find the alarm and use it quickly and effectively?
- Most alarms will emit a very loud noise but will passers-by take any notice? Some can sound like a car alarm and can be ignored.
- A load shout or shriek can be more effective.

If you slip, fall, injure or cut yourself, consider postponing the inspection until you have recovered. The smallest of cuts or bumps can get worse if they are not properly treated. Also, carefully evaluate any offer of first aid assistance from the occupier. Although they might offer this with the best of intentions, unless they are properly qualified they could make matters worse.

Property-specific safety issues

So far I have covered general aspects that will help you operate within a framework that protects your health and safety. Unfortunately, few guides give detailed guidance on the technicalities. For example, we all know that we should use ladders safely, but have we ever been on a properly structured training course? I suspect most surveyors developed their techniques 'on the job' resulting in well-established and possibly unsafe working practices. Therefore, I have provided below detailed advice for two of the most hazardous parts of a surveyor's job; loft inspections and surveys of derelict property.

Safety during loft inspections

Inspecting a roof space can present the greatest and most regular risk during a surveyor's day. Assembling the ladder in a restricted space can easily lead to an injury unless you take care. Although this sounds self-evident, I suspect that few surveyors have ever been trained how to use ladders properly and if you are new to the property sector, you may have no experience at all. Here are a few tips:

- Never stand on chairs, tables, or other furniture. If you need to inspect at a higher level, you should always use a ladder.
- Only use fitted loft ladders if you are satisfied they are safe – this is very important, as DIY-installed loft ladders can be very unstable. Take care when lowering the hatch mechanism, because the ladders can often slide down unexpectedly. Make sure you fully extend and lock the ladder into position, and test stability by putting your weight on a lower rung and shaking it. If you feel that the loft ladder is not safe, it is unlikely that you will be able to position your own ladder to get into the loft. This may well limit your inspection and you should state this clearly in your report.

Important tip: when you remove the loft hatch, dust and debris will shower down on both you and the landing carpet. In anticipation of this, put a dustsheet down under the hatch before you assemble your ladder. Leaving the homeowner with a lot of dust and dirt to clear up is not a positive start, or end, to your relationship! Additionally, be careful with the rubber 'end protectors' of the ladders as they can leave large black marks on both floors and walls. Many surveyors prevent this by temporarily fixing a variety of barriers over the ladder ends; including plastic bags, towels, and even old gloves. Whatever your choice, make sure they are safe.

Most surveyors use sectional aluminium ladders consisting of four or five sections that extend to between 3m and 4m in length (see page 13). Telescopic ladders are increasingly common, as many surveyors claim they are easier to use. Whatever the type, the ladders should conform to the appropriate standards; it should be complete; and you should follow the manufacturer's instructions.

There are a few important rules about using ladders:

- Each time the ladder is used, you must check it is in good condition and has not been damaged, especially if other people may have been using it. All the sections should be properly slotted together and all the bolts and nuts hand tightened.
- Safely position the ladder using a 1:4 ratio, i.e. set the ladder at 25 per cent of its height away from the wall.

It is often difficult to assemble a sectional ladder on a small landing – three sections can fail to reach the hatch, but there is not enough height to fit four sections together with the hatch still in place. You will face the dilemma of how to open the hatch itself. Here are two options:

- Prop the ladder against an adjacent wall, climb the ladder so you can remove the loft access hatch and set it to one side within the loft space. Caution: some loft hatches can be very heavy, partly painted or securely fixed shut. You will have to evaluate the risk and if you think it is unsafe, record the circumstances in your risk assessment section of the site notes. Climb down and add an additional section to the ladder so it extends well into the loft space. To be safe, there should be at least three rungs (preferably 1m) extending above the top of the ceiling joists.
- Alternatively, fully extend the ladder and use it to push the loft access hatch upwards just enough to allow the ladder to extend into the roof space. Once the ladder is securely in position, you can climb up and simply lay the hatch to one side. The drawback to this method is that unless you are able to balance the hatch on the ends of the ladder, you might easily dislodge it, and it may fall towards you.

The choice will depend on which you think is the safest.

Other considerations:

- Place the ladder on a firm and non-slip surface. Mats and laminate flooring are particularly dangerous.
- Make sure you wear suitable footwear that gives a firm grip on the rungs.
- Before climbing the ladder, make sure that the underside of your footwear is free from soil, etc. otherwise they may slip off the rungs.
- When on a ladder, never over-reach in an effort to inspect a specific feature. Get down and reset the ladder.
- Always keep 'three points of contact' with the ladder, for example both feet and one hand. Do not make notes on a ladder: writing on a clipboard means you have only two points of contact. Get down off the ladder and then make your notes.
- If possible, carry any equipment in your pockets when climbing a ladder. If you are carrying a torch, tape, and clipboard, it will be difficult to steady yourself and it will be easy to drop one of them on someone's head!

If you have to get off the ladder in the loft space, carry out a visual risk assessment before you leave the safety of your ladder. Look for the following:

- Can you see the top of the ceiling joists? If they are covered with insulation or stored items, you should not enter the roof space.
- Is there any boarding over the joists? If yes, will it take your weight and has the owner properly fixed it? Many owners put old wardrobe doors and other assorted panels down in place of properly fixed flooring materials.
- Are there signs of dangerous materials?
- There can be a variety of animals and insects living in the loft. There are few experiences in life that can match the adrenalin 'rush' of disturbing 15 roosting pigeons in a dark roof space. Although this sounds amusing, it can easily cause a serious fall. Other signs to look out for are flies, wasps, squirrels and even rats and mice. The droppings can also present a health risk.
- You should be alert to the presence of bats. Although they pose no serious health risk (do not worry, they will not bite your neck) they are protected by the *Wildlife and Countryside Act* 1981 and should not be disturbed. If you do discover bats, end your inspection, carefully and quietly withdraw from the loft and inform the owner of their presence.
- Because lofts are often dirty you should wear:
  - disposable gloves;
  - overalls; and
  - a face mask.
- When measuring the depth of the insulation, just lift a small corner. Do not roll large areas back as this will disturb dust, which could irritate your throat and skin.

Derelict and empty properties

A percentage of the properties you inspect will be empty, and this category includes: buy-to-let properties that are not tenanted; family homes where the occupants have moved out before they have sold the house; and homes where the owners have died.

If the property is empty, you need to clarify how you are going to gain access to the property. Getting the answers to the following questions may help:

- Where are the keys and how long can you have them for?
- Are there any security alarms and who keeps the alarm codes?
- Are the services activated and if not, will anyone be able to turn them on?
- Do you have permission to open windows, access hatches, etc?
- Who will answer all the pre-inspection enquiries (discussed below)?

If someone is meeting you at the property to open it up, make sure you ask them for proof of identity. This may appear a little formal, but it is important to make sure they are who they say they are so that any potential risks are minimised.

Being alone in an empty dwelling increases the level of risk. In addition to the general precautions identified earlier in this chapter, be aware of risks specific to empty properties:

- When you enter the empty building, carry out a cursory inspection of all rooms while loudly announcing your presence. The purpose is to make sure you are really are alone in the property. Make sure you look in all spaces including basements, stores and large cupboards.
- If you are on your own, lock the external doors when you are inside and keep the keys with you. Some commentators do not agree with this approach because it might hinder access if you have an accident. However, I would prefer this to a stranger casually wandering into the house when I had my head in a loft.
- If you discover an unauthorised person(s), briefly and calmly explain who you are, why you are there and that you are leaving right away. If the agent has told you it is empty then that is how you should find it. You should immediately phone the person responsible for the property and ask them to contact you when it is safe to inspect.
- The same approach is appropriate if you discover signs of unauthorised occupation and/or the property is unsecured. Typical examples would include a broken window or half open door and evidence that someone has been staying there (for example sleeping bags, open food tins, and so on). In this case, leave the property immediately and notify the person responsible.
- Properties left empty for long periods of time can become neglected allowing defects to develop quickly. Stagnant conditions can allow dormant wood rot to weaken timber floors and roof leaks will cause ceilings to collapse. Vandalism will often result in broken glass and sharp edges throughout the dwelling and drug-using squatters may leave contaminated syringes. Take the following precautions:
  - dress in robust and durable clothing including footwear that is suited to the conditions;
  - ensure you have a powerful torch with you, as many rooms may have their windows boarded over and will be in virtual darkness;
  - electric and gas services in empty properties can pose a danger especially if damaged through vandalism, so look for dangling wires, bare cables and a smell of gas;
  - empty properties can be cold during the winter.

I realise this final point is stating the obvious, but standing in sub-zero temperatures can be more than uncomfortable; it can be bad for your health. A pair of 'finger mittens' that allow you to use your pen but help prevent your hands becoming cold are invaluable. If you are 'follicularly challenged', a woollen hat is just the job. A flask of tea or coffee and sandwiches can help maintain concentration when the temperature drops.

LIAISING WITH THE OCCUPIERS

Before the inspection, you should get as much information from the seller or person in charge of the property as possible. The exact process and timing of this will depend on the circumstances, but the more you get, the better informed you will be.

The owners of all homes put on the market must produce a Home Information Pack (HIP) and despite the controversy (and I find it very hard to admit this) parts of it can be useful. For example, the Property Information Questionnaire (PIQ) includes many of the questions I would ask of the seller. Information usually includes: listed building/conservation area; parking arrangements; any storm or fire damage: flooding and flood risk; dry rot, wet rot, and damp together with any guarantees; electrical and heating checks; types of services; alterations/replacement windows and associated permissions; access/rights of way; and leasehold information.

The PIQ is a mandatory part of the HIP so if you can get a look (or even a copy) of the questionnaire, it can save you a lot of time. However – and this is very important – you should not accept the information without question. There is no substitute for collecting and verifying your own information. I would recommend that you still ask the seller whether they stand by the information. The practice notes are more explicit. It states that '... The surveyor should obtain a copy of the Property Information Questionnaire' [paragraph 1.5, page 4].

Using seller questionnaires

There are several reasons why I recommend developing your own questionnaire to give to sellers, rather than relying on the PIQ, not least of which is that for various reasons the information in the PIQ may no longer be accurate.

The process of developing and implementing your own questionnaire will largely depend on how you are instructed and the procedures that you prefer to follow.

For example, where there is time, you can send a questionnaire by post. This can save time on the day and it gives the owner time to locate all the necessary information (guarantees, certificates, etc.) at their leisure. However, there are a number of issues to resolve:

- Your questionnaire has to be written and presented in a fashion that suits a layperson reading it independently. In other words, it has to be understandable. Some questionnaires I have seen are very technical documents destined for the rubbish bin. Additionally, literacy levels do vary and some people are not confident at providing a written response. Others use English as a second language.
- If the owner has completed it, you still have to read the responses during your time at the property because many of the answers can give an insight into condition and will have to be followed up.

Alternatively, if you cannot send the questionnaire in advance, it could be completed at the time of inspection through one of two methods:

- Give the questionnaire to the occupier at the start and ask them to complete it while you are inspecting the property. This gives them some time to complete the form, find all the necessary information and it keeps them busy (so they do not follow you around). Although you must read through the completed form at the end to make sure you have the necessary information, this method still relies on the form being understandable.
- Sit down with the seller as soon as you arrive and go through each question verbally. This allows you to explain and discuss the topics in more depth, gives you a good insight into the property, and establishes a (hopefully) positive relationship with the person.

It is this last approach that I prefer. Whatever your choice, I have included a typical 'Sellers' questionnaire' in Appendix 2 that can be used alongside the site notes package.

Building a relationship with the seller – on arrival

A positive relationship with the seller can reveal a lot of useful information about the property. It is surprising how forthcoming a proud owner can be, even about the problems in their homes. Conversely, an over-attentive seller can get in the way of an inspection, making it difficult to approach the task in a methodical and systematic way. I have outlined a few tips below. To many, this may appear elementary but customer feedback often reveals how uncommunicative surveyors can be, so I will risk giving offence (for which I apologise) to set down a systematic procedure:

- Arrive on time with proof of your identity.
- Explain the purpose of the inspection – not all sellers understand the complexities of house sale and purchase procedures.
- Outline your approach to the inspection – how long it will take, what rooms and spaces you will need to inspect, etc.
- Deal with the seller's questionnaire (see discussion above).
- Ask the seller to show you around the property (maximum five minutes). This introduces you to the property, allows you to ask gentle but probing questions, and make arrangements for inspecting occupied areas, for example bedrooms with sleeping occupants, etc.
- Ask the seller to open any hatches and access panels that they would open during the normal use of the property – this will not only save you time, it will help you carry out a more thorough inspection.
- Ask the seller if there are any defective elements that may be damaged if operated or checked, such as windows that do not shut, taps that cannot be operated.

Once the initial and brief walkabout is complete, politely inform the seller that you want to carry out the rest of the inspection on your own.

Dealing with seller enquiries

Whatever the circumstances, many sellers will be very interested in what you thought of their home and what you might include in your report. This is understandable as they will be worried about the impact it will have on the sale. This can be a difficult discussion because you are there working for the buyer to whom you owe the primary duty. You must avoid being drawn into discussing controversial issues you may have just spotted. One of the best explanations is to say that you cannot come to instant judgements and you will have to review your notes back at the office before you come to any firm conclusions. It will help everyone if you stay as neutral as possible.

3. The Inspection Procedure

HOW FAR DO WE GO?

The *RICS HomeBuyer Service 2009 3rd Edition Practice Notes* and the explanatory text printed in the actual report describe the inspection. The new HBR is specifically designed to conform to the current 'level two' benchmark for surveys and so the scope of the inspection is the same as the HSV.

Rather than repeat this guidance, I have picked out 'themes' that regularly create discussions in my training seminars.

I think these themes are controversial for two reasons:

- They are the inevitable 'grey areas' where we have to exercise our professional judgement – the classic 'how far do I go' questions that are always useful to debate.

- After years of carrying out the same type of service, something that I call 'mission drift' can creep in. This is where surveyors imperceptibly adjust what they do. For some, this will result in an inspection routine that falls well below the current benchmark while others exceed it by a considerable margin.

Both of these extremes are inappropriate.

If we are to achieve consistency between surveyors across the country and protect the reputation of this new product, we need to articulate 'what is a reasonable inspection?' I hope this discussion helps.

Note: At the beginning of each theme, where appropriate, I have shown relevant and direct quotes from section B of the report, pages 21–29 (shown in *italics*, with page references in square parentheses), and the description of the RICS HomeBuyer Service (the small print at the back of the report).

Theme 1: *'The surveyor ... will examine floor surfaces and under-floor spaces so far as there is access to these ... '* [page 35] *'... but does not force or open up the fabric. This means that they do not take up carpets, floor coverings or floorboards ... remove secured panels ... '* [page 24]

The practice notes also states [page 27] *'In England and Wales, where the sub-floor access is by trap, the inspection is restricted to the components and over-site visible from the trap on an "inverted head and shoulder" basis'*.

This is very comprehensive advice and although I generally agree, the application of sensible judgement is required. For example:

- Floor coverings are not lifted, but if they are loose (by this I mean loose rugs, fitted carpets very easily lifted with the fingers, etc.), I think we should lift an accessible corner, as this can be very revealing.
- We should not remove secured panels. On a 'level two' inspection, we should not waste time undoing cupped screws to get into a roof space. However, we could ask the owner to do it if they were willing and able.
- The inspection of under floor spaces does pose the occasional difficulty. If there is a purpose-provided access panel in a timber ground floor that is easy to remove or to open, then we should carry out an 'inverted head and shoulder' inspection. In other words, you should stick your head down. Using a torch and an inspection mirror can extend your field of view. We should not go into the under floor void unless the opening is large enough, the void is deep enough and there are fixed steps safe to use.

Figures 3.1 and 3.2 illustrate two typical examples.

Figure 3.1: This photograph shows the surface of a timber floor just inside the under-stairs cupboard of an older home. The hallway floor is solid and covered with an original tiled finish. The loose carpeting was easy to lift and to set aside, showing loose floorboards beneath. These were not nailed or fixed in anyway and could be removed with a standard screw driver without any difficulty

Figure 3.2: Here, the floorboards have been removed, revealing two useful pieces of information: a section of the rising main in lead piping (it changes into copper before it becomes visible); and the area of suspended timber floor beneath the stairs surrounded by a solid floor, is not ventilated and vulnerable to dampness and rot. In my view, this should be lifted during an HBR

It is different in Scotland

A different approach has developed in Scotland. Surveyors north of the border regularly carry out physical inspections of the sub-floor area if the surveyor '… deems it safe and reasonable to do so'. The usual measure is there has to be at least 1m clearance between the underside of the floor joists and the sub-floor surface (sometimes called a solum).

Figure 3.3: Although the contents of an under-stairs cupboard should not be moved, the stored items conceal both the gas meter and the consumer unit for the electrical system. Applying the 'follow the trail' principle, enough items need to be moved to enable an inspection of these important features

Theme 2: '[The surveyor does not] … *undo electrical fittings …*' [page 24]

You would think the majority of practitioners would never dream of doing this – but some do. Not only does it take surveyors beyond their level of competence, it is dangerous. Do not do it!

Theme 3: '[The surveyor does not] … *move furniture, remove contents of cupboards …*' [page 24]

The furnishings in most occupied dwellings will limit the extent of an inspection and it is clear that the HBR is a non-invasive inspection. However, most clients would be disappointed if you refused to move small and mobile items, especially when the property is empty when they move in. I think it is reasonable to move the following:

- Easy chairs, sofas, beds, and the like, where there is enough space and they are on castors, making it easy for one person to push aside.
- Small coffee tables, magazine racks, dining room chairs, standard lamps, waste baskets, and so on.

With cupboards, spending time taking out (and putting back in) boxes, vacuum cleaners, buckets, and so on, is clearly beyond the remit of the HBR. However, carefully moving aside a few cleaning products and old

Figure 3.4: These few items under a sink should be moved during an HBR. They are easy to move and quick to replace

dusters so you can see the services under the sink is something we should consider. Figures 3.3 and 3.4 show some typical examples. In all cases, the owner's permission must be obtained before anything is moved; some people can be very particular.

Theme 4: '[The surveyor] ... *will inspect the roof structure from inside the roof space if there is access (although they will not move or lift insulation material, stored goods or other contents).*' [page 35]

'*The surveyor ... may use a ladder for flat roofs and hatches no more than three metres above level ground (outside) or floor surfaces (inside) if it is safe to do so.*' [page 24]

I have already discussed the health and safety implications of using a ladder to inspect the roof space and so please refer to page 22 for my definition of the term 'safe to do so'. Assuming that it is safe to get into the loft, on a 'level two' inspection, I think we should be prepared to get off the ladder to carry out a 'general inspection' of the roof space that would include:

- a close inspection, including taking moisture meter readings of a small but representative sample of timber components especially those vulnerable to deterioration;
- a visual inspection of any tanks and pipes, etc.;
- a visual inspection of the junction of the chimney and the roof covering;
- lifting just one small corner of an insulation roll so that you can measure its thickness and help identify the nature of the ceiling (although you should not roll back insulation to inspect the back of the ceiling).

We should not have to rearrange stored items so we can get around, and health and safety considerations will quickly limit our activities. The most important point I want to make here is to try to 'fix' the 'level two' loft inspections. I hope this helps:

- The 'level one' inspection (mortgage valuation) is a 'head and shoulders' only. The surveyor does not get off the ladder.
- The 'level two' inspection goes beyond a head and shoulders and the surveyor should get off the ladder to inspect the features described above.
- The 'level three' inspection (the building survey) is a detailed inspection of all components. The surveyor will need to wear overalls, be prepared to crawl around a bit and get dirty.

Figures 3.5–3.7 give a number of typical examples of loft spaces and Figures 3.8 and 3.9 show a flat roof.

Theme 5: '*The surveyor will not carry out specialist tests, or test or assess the efficiency of electrical, gas, plumbing, heating, or drainage installations ...*'

Most surveyors would agree that the services are not operated or tested in any way. I have had so many arguments in training seminars about whether turning on a tap is a 'test' I have almost lost the will to live. On a more serious note, this is another critical difference between a level two and three inspection:

Figure 3.5: This is a clear roof space, the backs of the ceiling joists are visible, there are no obvious tripping hazards, the joists look strong enough and the purlins and rafter provide good hand holds. I would go into this roof space

Figure 3.6: This roof space has been insulated with blown thermal insulation made out of recycled paper. The ceiling joists are concealed and it would be unsafe to proceed. I would NOT go into this roof space

Figure 3.7: I wish all roof spaces were like this – plenty of head room, no stored items and substantial chipboard properly fixed to the ceiling joists. But always make sure the flooring is fixed and not loosely laid

Figure 3.8: This property has a single-storey extension to the rear. Usually, this can be inspected from an overlooking window

Figure 3.9: This is the view of the roof in Figure 3.8 from the closest window. Can you see enough of the flat roof to make a decision or do you need to look at it from the top of your ladder? This will always be your decision but a 'not inspected' or a 'restricted inspection' are not appropriate as the flat roof is less than three metres off the ground

- 'level two' – no tests are carried out, a visual inspection only;
- 'level three' – where they are switched on, services should be tested through normal operation. For example:
  - taps are run, toilets flushed;
  - the owner is asked to switch the heating on;
  - water is observed running through the inspection chambers (what I call the 'flush and run' technique).

At 'level two', the inspection chamber lids are lifted (if it is safe to do so) and you should visually inspect the chamber itself. There is no need to engineer water to run through it.

Inspection of flats

The inspection of flats is unchanged in the new HBR. The Description of the RICS HomeBuyer Service (DHS) says the inspection includes:

- the assessment of the general condition of the outside surfaces of the building;
- the same approach for shared hallways, staircases, and so on;
- inspection of the roof space but only if you can get into it from within the flat itself.

However, the drain covers will not be lifted and the lifts, fire alarms or security systems will not be assessed. This broadly matches the previous editions of the HomeBuyer.

Surveyors should also be familiar with the 'Leasehold Properties' advisory leaflet (LPA), which should be inserted into the HBR pack for flats. This describes the assumptions on which the report and market valuation are based; important factors that can influence the inspection and report.

Although I do not want to reinvent the wheel, I think it is worth clarifying this inspection procedures for flats, for the following reasons:

- Over the past few years, thousands of flats have been built across the country, so all surveyors will be asked to inspect flats from time to time. They are no longer primarily restricted to metropolitan areas.
- Flat owners may have to contribute to the upkeep of the block/estate, through the terms of the lease or the common hold. If the block is in a poor condition, the bill could run into thousands and so influence their decision to buy.
- It is important to collect the correct data during the inspection, so that you can develop a consistent approach to rating and reporting for both flats and houses.

The extent of the inspection

In my view, this will depend on the size of the development. If the flat is one of just a few dwellings squeezed into a converted former single residence, then your inspection will not be that different from a house and typically involve an inspection of:

- the shared hallway and staircase from the top to bottom (including any shared meter or other cupboards);
- a loft inspection, but only where the access hatch is within the subject flat;

Following the trail

When outlining the focus and limitation of the HBR, the practice notes state that '... *where there is "a trail of suspicion" the surveyor "must take reasonable steps to follow the trail"*' [paragraph 2.4, page 6].

The practice notes then refer to *Roberts v Hampson and Co* (1989), which established this well-known precedent. In that case, Judge Kennedy stated: 'if there is specific ground for suspicion and the trail of suspicion leads further behind furniture or under carpets, the surveyor must take reasonable steps to follow the trail until he has all the information which it is reasonable for him to have before making his valuation.'

Although aspects of this opinion have been modified by others cases, it clearly indicates that surveyors must go beyond the scope of inspection where appropriate. This does not mean that we have to cancel the rest of the day's appointments and go to get the floorboard saw from the car, but it does mean we have to be flexible. To try to illustrate this, I have include an example in Figures 3.10 and 3.11.

Figure 3.10: The skirting to the inside of a bay window: fitted carpet is in place but the uneven surface to the paint finish clearly indicates the timber is rotten. This creates a 'trail of suspicion' and the next question is 'How far does this rot go?'

Figure 3.11: Here is the skirting in Figure 3.10 after I had pulled a corner of the carpet back. This was easily achieved because the carpet was not on grippers or stuck down. The floorboards were damp, but were not affected by the rot. Further investigations were still required to make sure the floor joists were problem free but following the trail and going a little beyond the normal inspection was justified

- the external walls, roof, chimneys, gutters, downpipes, windows and doors from shared parts of the grounds, those parts belonging to the subject flat, and/or any adjoining public space.

The differences are that the inspection chambers are not lifted, and the inside of the other flats cannot be seen.

In terms of the development's size, two assumptions in the LPA are important:

- Where there are six or more flats, you should assume that the development is managed by the freeholder or a managing agent. Less than this number, the assumption is that it is the freeholder alone. Therefore you must establish the number of flats in the development.
- If there is more than one block in the development, the lease terms apply (except for upkeep of common roads, paths, grounds and services) only to the block the property is in.

This last point is important. Where the flat is in a block on an estate that contains a number of other blocks, do we simply put on our 'blinkers' and ignore the others? Despite the clarity of this particular assumption in the LPA, I think it is important to at least make a note of the condition of the other blocks in your site notes, especially where the other blocks are worse than the one at which you are looking.

For the rest of the inspection, you will have to make a judgement about how far you go. Here are my thoughts:

- In the shared internal areas, be prepared to walk around a little to get a 'feel' for its condition. At the very least, inspect the floor the flat is on and the one above and below if accessible and easy to do.
- Externally, inspect all elevations of the block from accessible shared areas, but restrict your inspection of other parts of the estate to what you can see from the immediate area outside the subject block.
- Although the practice notes exclude assessment of the lifts, what happens if you use the lift to get to the flat and you cannot fail to notice its poor and sorry state? Such a case is a matter of judgement,

but in most cases you should mention this in G7 Common Services (see below), providing you make it clear that this does not constitute a full assessment.

There are a number of other issues raised in the LPA you need to look out for:

- Right of access over all shared roads, corridors, stairways and so on, and the right to use shared grounds, parking areas and other facilities, is assumed. Therefore, if you notice any restrictions on use (for example parking or garden areas with locked gates, etc.), put that in your site notes and raise the issue under section I of the report;
- The HBR assumes there is no current dispute, claim or lawsuit relating to the lease, and as surveyors do not carry out any legal searches or investigations, this assumption will normally stand. However, I find standing in the middle of an adjacent common area with a clipboard never fails to bring out the neighbours. Many are happy to let you know about the relative merits of the freeholder and/or managing agent. Anecdotal information should always be treated with caution but combined with other information, it can help you form an overall opinion of life on the estate.
- Are the other flats occupied by owners or tenants? Your 'chat' with neighbours could help here otherwise look out for 'to let' boards, etc.
- The shared and common areas of the estate are subject to the 'Control of Asbestos Regulations 2006' and the legal adviser should be asking for evidence that the appropriate measures are in place. However, because you are the 'eyes and ears', keep a look out for any obvious and potential asbestos-containing materials such as 'asbestolux' boarding, textured finishes in common areas, and so on. Where spotted, you should emphasise these in the report and section I;
- Although the legal adviser will be asking about cleaning and maintenance arrangements, what is the evidence on the ground (sometimes literally)? You should include any concerns in your report.

Once you have collected the data, you can then allocate appropriate condition ratings (chapter 5) and write you report (chapter 6).

INSPECTION PROCEDURE

Most experienced surveyors will have their own approach to carrying out an inspection and I think it is beyond the scope of this book to cover this in detail; and I have written about the process elsewhere (Parnham, P. and Rispin, C., *Residential Property Appraisal*, Spon Press, 2001). The main rule is that your approach must be methodical and systematic, following the same routine in every inspection. This can help enhance the impression of competence if challenged in court.

Figure 3.12: I took this photo from the top of a small hill within the grounds of the estate containing the blocks. Should we be hiking around the grounds in an effort to catch a glimpse of the roof? This will always depend on the circumstances, but I think we should use all accessible vantage points within the estate grounds and adjacent public spaces if required

Site notes

Whatever the methodology you adopt, it is essential that the information about a property is recorded accurately and clearly. In Appendix 1, I have included three different versions of my site notes each designed to accommodate different preferences – Appendix 1a Site notes; Appendix 1b Site notes (room by room supplement); and Appendix 1c Site notes (for dictation). The rationale for these different versions is explained below.

In my experience, many surveyors do not adequately record what they see during the survey. When asked why, I have been given a number of reasons including:

- 'I recorded just the most important information, I can remember the rest';

Figure 3.13: The subject property is in the centre of the picture on the first floor. Where the balconies are open, it is easy to make a general visual assessment of the other floors. If the balconies were enclosed, this would not be possible, and on an HBR I think you should restrict the inspection to the corridor that includes the flat. You should make this restriction clear in your report

- 'I intended to finish them off later' (but he never did);
- 'I used a mortgage valuation site notes form but wrote more';
- 'I took loads of photos and I wrote the report using them';
- 'I do not have enough time to make a complete set of notes'.

This is simply not good enough. Busy professionals cannot remember everything and without a complete record, it is impossible to produce a balanced, accurate report. Claims against surveyors often succeed because their competence is undermined in court by their poor site notes. Site notes are important because they:

- help you record everything that you see on site and in the vicinity;
- act as a checklist to ensure that you have included everything;
- give an opportunity to reflect on the survey from which you can produce the report; and
- provide evidence that you have undertaken the inspection in a logical and thorough manner and in doing so help show competence.

However, there are many different ways of achieving these objectives; we do not have to follow exactly the same methods. For example, some people record information visually by annotating drawings and sketches; while others are happier describing what they see in words. Some dictate and record their site observations, and a few have very personalised recording systems that are unintelligible to others but make sense to them.

A special word about dictation ...

Surveyors have to write a great deal, so many choose to record information on a handheld personal recorder. Whether it is a letter, memorandum, site observations or a full report, once completed a small tape or a digital file is usually passed to the office secretary and the typed draft later checked, amended and sent off to the recipient.

In respect of site notes, the recorded observations are normally typed so the document acts both as a permanent record and a memory aid that helps the surveyor properly reflect on the inspection as a whole.

This approach was originally recommended in the RICS HSV practice notes, and reinforced by the case of *Watts v Morrow* (1991) 2 EGLR 152, (1991) 4 All ER, 937, CA. Here, the court was critical of a surveyor who dictated the full report while on site. The judge considered the report unnecessarily long and lacking reflective thought. In other words, the surveyor did not record his observations; he skipped that stage and instead dictated the actual text of the report during his inspection. This may save time but, in my opinion, it is impossible to come to an instant view about a defect.

Consider, for example, an area of dampness at the base of a typical external wall. To diagnose the cause, moisture meter readings will have to be taken, the inside and external surfaces fully inspected, the carpet and floorboards lifted (if loose) and the external grounds levels assessed. Assimilating all this information and then dictating what is to go in the report while still at the property is simply impractical.

To avoid this problem, if you use a personal recorder, you should record what you see – focusing on the signs and symptoms. Once your observations are typed up, you can read them, account for all the relevant factors, and only then begin to dictate your report. Although this process is broadly accepted, my own anecdotal discussions with surveyors have revealed a number of persistent problems:

- Many find it difficult NOT to dictate their report during the inspection, as habits developed through years of experience prove hard to break.
- Most rely entirely on the spoken word and are reluctant to use any form of written site notes. This may be acceptable for mortgage valuations or other services of limited scope but not for 'level two' products. For these, a proportional sketch of the site and each floor is essential to the assessment of more complex problems such as structural alterations, movement problems and drainage issues, and so on.
- Few surveyors who dictate use an inspection checklist, and those who do rarely show they have 'checked' the items on their list. They prefer to follow 'well established routines' that have been '... followed for years'. Experience is obviously valuable, but what happens when they get distracted or are not feeling very well or it's simply getting towards lunchtime and concentration levels are falling? In my view, this is when things are missed and mistakes are made.

To summarise, I would propose the following procedure if you use a personal recorder:

- record only your site *observations* during the inspection, do not begin to dictate the report;
- use a paper-based site observations package for the site and floor plans;
- use a series of 'annotated' checklists to make sure you have looked at everything;
- use the sketches, printed site observations, and any photographs as the basis for dictating your report.

If these different methods meet the standards a court would consider appropriate, it is unwise to change them. Instead, you should review your own approach and adjust appropriately. This is particularly important for 'more experienced' (in other words, older) colleagues who may have developed their methodology when the HSV was first developed and not changed it since. Expectations change and it is important to make sure we always keep in step. Here are just a few suitable approaches I have encountered:

- handwritten notes and varying amounts of sketches and drawings on pre-printed sheets;
- dictated and recorded observations that are later typed as a permanent record;
- highly structured inspection forms that use coding systems completed during the inspection on paper or, increasingly, using handheld personal digital assistants (PDAs).

Site notes packages

The Site notes 'packages' presented in Appendix 1a–c should really be called 'observation packages' because 'site notes' suggests written text and this is too limited for what they are – but we will stick with their more conventional name.

The packages are designed to suit several different data recording methods:

- 1a is for handwritten site observations following an elemental structure;
- 1b is for handwritten site observations using a partial 'room by room' structure;
- 1c is for dictated site observations with handwritten sketch plans and checklists.

The 'Sellers' Questionnaire' (Appendix 2) and the 'Market Valuation Analysis' (Appendix 3) are designed to be used alongside any of the pro formas in Appendix 1.

I followed a number of general principles when designing these forms and the following are common to all variants:

- I have tried to avoid too many 'tick boxes'. The tick box approach is common for a 'level one' inspection and allows little flexibility when non-standard features are encountered. Wherever possible, I have created a simple text box to allow for additional notes, sketches, etc.
- A checklist has been included for each element and this has two important functions:
  - to act as a memory aid to make sure you look at everything (I have tried to include the most important features but if I have missed something, please include your own items on your master copy);
  - to show you have looked at these things during your inspection ideally, you ought to tick off each item, but I find that too time consuming and instead, when I have checked an element, I simply draw a line through the whole lot.
- In the checklist, I have included only those features that you should be looking at during a 'level two' survey. On seeing these lists, some surveyors have said they describe a building survey. I do not agree; the checklist for a 'level three' survey is even longer. This goes back to my central assertion (see page 14): in the past many surveyors have not gone into enough depth on a 'level two' inspection and report.
- Due to the size and format of the text boxes, rather than leave white space at the bottom of some pages, I have included an 'other comments' box. These should act as an 'overspill' from other pages. There is also an additional overspill page at the end.
- Space for the property address is included at the top of each page just in case the individual pages become separated.

These different options are explained below. If one does not suit you, produce a variation of your own!

Handwritten site observations (elemental structure) (Appendix 1a)

This package uses the final report as its structure. It is split into several distinct parts:

- General details (pages 1–3) including details about the surveyor, the inspection itself and most of the information required for sections A to D inclusive.
- Site and floor plan sketches (pages 4–6). Three full pages have been included, but if the property is larger or you prefer to use more sketches, simply photocopy more pages and add them to the pack. Each page is printed with a faint grid of squares to help you produce proportional sketches. You should add your own key to each set.
- Section E – Outside the property (pages 7–12). As the name suggests, this covers the external elements of the building.
- Section F – Inside the property (pages 13–18). This follows the internal elements of the report.
- Section G – the Services (pages 13–24). This section is for all the services covered in the new HBR.
- Section H – the Grounds and shared areas for flats. This section includes the new element of garages, other outbuildings and the grounds. This is an important section for flats where all external features should be included.

One of the main advantages of using Appendix 1a is that the information is collected in the same order as the report. This makes writing the report straightforward, because you do not have to search for the information. The main disadvantage is that the layout does not follow the natural method of inspection. For example, most surveyors will inspect a building space by space, then floor by floor. If you use

Appendix 1a, you will have to flip through the pages as you go through each room so that you can place a comment against the ceiling, then walls, floors and so on. Despite this inconvenience, I prefer this form, and it can be made easier to handle by printing the notes on one side only and connecting them with a good old fashioned Treasury Tag.

Handwritten site observations (room by room supplement) (Appendix 1b)

This version may better suit surveyors who prefer to collect the information room by room. Consequently, the pages for section F are different and the pro forma should include:

- General details (pages 1–3).
- Site and floor plan sketches (pages 4–6).
- Section E Outside the property (pages 7–12).
- Section F Inside the property (pages 13–15), which has been reduced to include:
  – 'Restriction on inspection' box and F1 roof space and structure as before (included because it will apply to most properties and is different in nature to the other functional spaces).
  – Room-by-room data collection sheet. You should take enough copies to cover the number of rooms in the property. I have designed it to suit a generic space because function specific headings would be too inflexible. There is little space for a full set of checklists so these have been included separately.
  – A checklist for elements F2–F8 inclusive. This should be used to make sure you have covered everything in each room (see pages 24 and 25).
  – The section F9 has been included for the same reasons as F1.
- Section G Services (pages 15–21). Although the services would be described in the 'room-by-room' pages, I think that property-wide services issues are so important they still need to be drawn together on their own sheet.
- Section H Grounds and shared areas for flats (pages 22–23).

This option will certainly make the recording of internal information follow a more natural pattern, but you will still have to flip back and forward through the pages when assembling the elemental report. Still, I guess you can do this in the comfort of your own office with a cup of tea by your side.

Dictated site observations with handwritten sketch plans and separate checklists (Appendix 1c)

I had better nail my colours to the mast: I find it impossible to dictate anything. I am unable to verbalise the content of the simplest of letters, let alone a complex report. However, not everyone suffers from a similar impediment and to many of my colleagues, past and present, dictating is as natural as writing.

When discussing the pros and cons of recording information in this way, one surveyor pointed out that in a given fixed period of time, more information can be recorded verbally than could be written down on paper. I found this argument persuasive and encouraged by a daughter more at ease with the modern world, I have now started using an MP3 personal recording device. Although it is still early days, the experience seems to be generally positive. The breakthrough came when I realised that I did not need to produce grammatically correct, well-balanced prose. Instead, all I needed to do was record what I see using my own unstructured, jargon-laced, vernacular language. The resulting MP3 file can be downloaded to my computer as a permanent record so I can review its contents at my leisure. In my view, this allows for the considered 'reflection' a court would expect to see.

However, despite these advantages, I think the general details should be handwritten (largely because there is not much to write) as well as the site and floor plan sketches. The rest is made up of checklists to follow when carrying out the inspection. Therefore this form includes:

- general details and checklists (page 1–3);
- site and floor plan sketches (pages 4–7);
- checklist for section E – Outside the property (pages 8–9);
- checklist for section F – Inside the property (pages 10–11);
- checklist for section G – the Services (pages 12–13);
- checklists for section H – the Grounds and shared areas for flats (page 13).

I hope you find these useful.

4. Understanding the Condition Ratings

Having co-written the *Home Inspector's Handbook*, it is difficult to write about condition ratings without launching into a cynical review of the reform of the home buying and selling process over the past 12 years. As this will be both tedious and bad for my blood pressure, I will simply side-step the past. If you want to know more about the origins of condition ratings, please buy a copy of the previously mentioned publication (see www.ricsbooks.com). Instead, I will discuss condition ratings in relation to the new HomeBuyer Report only.

WHY HAVE CONDITION RATINGS BEEN INTRODUCED?

The Homebuyer Survey and Valuation (HSV) did not include condition ratings, so why were they felt necessary for the new HomeBuyer Report (HBR)? Although other commentators may have different views, my explanation is as follows:

For years, surveyors have been categorising condition to help property owners understand the physical state of their buildings and to set priorities for their repair strategies. In 2007, RICS talked to consumers about what they wanted from chartered surveyors (see page 3 for a more detailed discussion). One of the findings was that consumers clearly wanted a brief 'bullet point' type summary of the report. Condition ratings were seen as a method of addressing that demand.

In December 2008, the Scottish Parliament introduced the Home Report. This mandatory package of information includes an energy efficiency report, a property questionnaire, and the 'Single Survey'. This is a condition and market valuation report produced by chartered surveyors that uses the condition rating system.

Although the original condition rating system for the HCR was not popular with surveyors, it is worth remembering that the process itself was developed *by* surveyors and has since been thoroughly tested in practice. As a consequence, condition ratings can be seen as a way of responding to public demand, while also aligning the HBR to developments north of the border. However, there is a fundamental difference between the HBR and the Home Report in Scotland: in Scotland both the buyer and the seller will be relying on the Home Report, so the condition ratings *have* to be applied objectively. There is no opportunity of explaining why the particular category was chosen. In the new HBR for England and Wales, this will not be the case. Although you will have to follow strict rules in establishing the rating, because it is a product solely for the buyer, you will have more flexibility in explaining your choice and to place that decision in context. To help you get into the right 'mindset' for condition rating in the HBR and make a break with the past, I will make a bold statement: **condition ratings are not the 'be-all and end-all'.**

The condition ratings are intended to add emphasis to the report; to highlight the most important issues. They are a tool that you can use to bring your client's attention to critical matters; they are not a universal truth to which we all must submit. They serve the same function as the 'F1s' and 'F2s' did in the HSV; they are there for emphasis.

Although the condition rating process in the HBR has clear rules, it should never lose sight of its main purpose: meeting the customers' needs. Have I made myself clear? Good, then I will continue.

DECONSTRUCTING THE DEFINITIONS

If you are familiar with the Home Condition Report, the condition rating definitions used in the HBR will be recognisable, although there are some subtle differences. Table 4.1 shows the ratings used in both the HBR itself and *The RICS HomeBuyer Service 2009 3rd Edition Practice Notes* [page 35].

Although the practice notes define a number of the terms used [practice notes, Appendix A, page 74]), a detailed explanation of relevant terminology will be helpful before I go on to describe the condition rating process in detail.

Table 4.1: Condition rating categories in the HBR

Condition rating	Definition	Additional comments
One	No repair is currently needed. The property must be maintained in the normal way.	This rating must be used where there are no indications of present or suspected defects that require the undertaking of a specific repair. Normal maintenance must not be treated as a repair for the purpose of the Homebuyer Report. The tests for this condition rating are: ● *Is there a repair or a replacement required to the element?* ● *Is the work required no more than normal maintenance?* If the answer to both is NO then the rating applies.
Two	Defects that need repairing or replacing but are not considered to be either serious or urgent. The property must be maintained in the normal way.	This rating must be used where repairs are required, but the defect is not considered to be serious or urgent. The tests to apply to the defective element are: ● Is it a 'serious defect' which compromises the structural integrity of the property? ● Is it 'urgent', where the repair cannot be deferred until the next anticipated period of regular maintenance without allowing the defect to develop into a structural failure, or where it could lead to a 'serious' defect in another element of the building. If the answer to both is NO then the rating applies.
Three	Defects which are serious and/or need to be repaired, replaced or investigated urgently.	This rating must be used where the defect is of a serious nature or where immediate repairs are required, or where the surveyor feels unable to reach the necessary conclusion with reasonable confidence. The tests of a 'serious defect' are: ● Does the defect compromise the structural integrity of the property? ● Does the defect impair the intended function of the building element? The tests for 'urgent' are: ● If the defect is not repaired/remedied immediately, will it cause structural failure or serious defects in other building elements? ● Does the defect present a health and/or safety threat?
NI	Not inspected	This rating must be used when it is not possible to inspect any parts of the dwelling usually covered. If the surveyor is concerned about these parts advice must be given about further investigations that are needed.

Defect

One definition of a 'defect' (which was used during the early stages of development of the Home Condition Report, but is not given in the practice notes) is: '… damage, imperfection or fault, which impairs, restricts or limits the functionality, normal maintenance requirements and expected lifespan of similar building elements'.

It is important to note that the new practice notes do not give a definition of 'deficiency' (unlike the Home Condition Report, which did offer a definition of the term). The following example might help to clarify the difference between a 'defect' and a 'deficiency', and the implications for the condition rating.

A 100-year-old house still has the original windows, which would fall below current building regulations standards. To apply a condition rating, ask yourself whether the windows need repairing ('Is paint peeling? Are parts of the wood rotten?'). If the answer is no, then there is no 'defect' and a condition rating of one could be applied. If work is needed, then rate accordingly. However, if the window is fixed and never had an opening casement – even though it is in a satisfactory state of repair – it cannot provide the ventilation that is needed, or the means of escape in case of fire. This is clearly a 'deficiency'.

In other words, the condition rating will not be affected by a building element that does not meet current standards because of its design, construction or manufacture. This has important implications when assessing health and safety hazards (see page 59).

Normal maintenance

The practice notes define this as '… work of a recurring nature which certain building elements routinely require in order to preserve their integrity and functionality.' [practice notes, Appendix A, page 75].

Normal maintenance may also include 'planned preventative maintenance', i.e. work that is carried out at predetermined intervals or to other prescribed criteria and intended to reduce the likelihood of an item not meeting an acceptable condition. Planned preventative maintenance usually has two components:

- servicing – scheduled work at regular intervals to prolong life and prevent breakdowns (such as annual clearing out of gutters and drainage gullies, yearly servicing of boilers and fires, repainting of wood and metal surfaces, etc.); and
- repairs and replacements – planned programmes based on observed condition, to repair or replace components immediately before anticipated failure (for example, replacing all boilers on a housing estate after 15 years regardless of their condition; an approach that is more common with larger scale, non-domestic property owners).

Because home owners would see planned programmes of maintenance as a 'repair' rather than 'normal maintenance', I will not consider it further.

For older buildings, distinguishing between normal maintenance and a repair is not always straightforward, because the performance of an older building element will not match the modern equivalent. Consider, for example, a solid brick wall that is 225mm thick. This is common in many parts of the country and it means that the wall is more susceptible to moisture penetration than a modern cavity wall, and it has a higher U-value, making it vulnerable to condensation. In training seminars, surveyors often 'punish' solid walls by placing them in a higher repair category. This is inappropriate for a number of reasons:

- Speculating that solid walls will be more vulnerable to defects in the future is a subjective value judgment, which will give older buildings an unreasonably negative image.
- The condition at the time of the inspection is a very important factor. If the wall is not letting water through or allowing condensation to form at the time you look at it, you should judge it as 'satisfactory'.
- Placing the performance of a solid wall against a cavity wall will skew the final rating decision. You should always compare 'like with like'.

Just to prove that life is never simple, it is important to understand that solid and cavity walls are different when they begin to deteriorate. For example, a gutter leak running down the face of a poorly pointed solid wall is more likely to lead to internal defects and require a more urgent response than a cavity wall suffering from the same problem. In this way, a solid wall will often change to a higher category more easily than an equivalent cavity wall.

Another challenging example is an older natural slate roof. 'Normal maintenance' for a 100-year-old roof covering could, in reality, involve a considerable amount of repair work. Corroded fixing nails and deteriorated slates will often mean that large numbers of slates need clipping or replacing after most winters. Surveyors may appreciate that this is 'normal', but could we say the same for the general public? I doubt it. To them, normal maintenance for a roof probably means replacing the odd slate every few years; not 10 to 20 every year. Although this can be covered in the report (see page 83), in some cases you will need to reflect this in the rating.

Serious defects

The practice notes go no further than: 'Those defects which compromise the structural integrity of the property and/or impair the intended function of the building element.' But what does it mean by structural integrity and function of the building? To ensure you adopt a consistent approach, it is important to be clear. Here are my thoughts:

- 'Compromise' or 'impair the intended function of the building element' – Every building element will have a number of functions. For example, windows have a primary function to let in daylight, with secondary functions to: be strong enough to resist the forces applied on it; allow for ventilation; keep the weather out; be durable and free from maintenance; provide for escape in the case of fire; resist the passage of sound and heat; and provide security. Not all of these secondary functions are equally important, and you will have to make a judgement. For example, one or two sticking casements would have little impact on the function of the element, whereas if all were stuck fast because of swollen and rotten timber, the rooms could not be ventilated, people could not escape during a fire and parts of the window could possibly fall onto passers by below. Problems of this magnitude would seriously impair the function of the element.
- 'Compromise the structural integrity' of the building element – The phrase 'structural integrity' conjures up an image of unstable walls or distorted roofs that are on the verge of collapse. Yet few buildings (or even parts of them) actually fall down. The dictionary meaning of 'integrity' is 'the

condition of having no part or element taken away or lacking; completeness' (Shorter Oxford Dictionary, 2002). Therefore, for a defect to 'compromise integrity', it would need to have a pronounced effect on an element, leaving part of it potentially unstable. It should not be limited to 'dangerous structures' but reserved for those parts of the building that are heading in that direction. Examples could include: a ceiling that is close to collapse (not simply cracked); part of a wall that has distorted so much it is unable to properly support weight from above; several roof components that have cracked or split and resulted in a very uneven roof slope; and a floor that vibrates excessively during a 'drop heel' test and has cracked the ceilings below.

It is impossible to give an exhaustive list and you will have to make judgements on a case-by-case basis.

Urgent defects

The definition in the practice notes is: 'Defects which if not repaired/remedied immediately, will cause structural failure or serious defects in other building elements ... ' Deconstructing this definition, there are three components:

- Repaired/remedied immediately – The dictionary definition of 'immediate' is 'taking action without delay'. This clearly suggests quick action is required to prevent matters from getting worse. In relation to the condition of buildings, this is always going to be a relative term. Consider a few examples: a gas leak will need repairing within a few hours to avoid an explosion; a bad roof leak will require swift attention (say within 24 hours) to prevent further damage to ceilings below; and a foul drain that is blocked and resulting in sewage overflowing into the garden will need clearing within a day or so. All of these would be considered as urgent. However, you should not be speculative. For example, not all minor gutter leaks will result in rampant dry rot after a few weeks. Most buildings are generally durable, resilient structures and few defects develop so swiftly (but I bet you know of a couple of exceptions!).
- Structural failure or serious defects in other building elements – The definition of structural failure is similar to 'structural integrity' and to be 'serious' clearly suggests that the defect must have an impact on other parts of the building. Examples include: a leaking gutter that has damaged the plaster to the walls and caused wet and dry rot in the adjacent timbers; building movement that has prevented a number of doors and windows from opening and closing, leaving a small portion of the wall as unstable; and dry rot that has caused a number of floor joists to collapse. To come within this definition, the consequences of the defect have to be serious and the deterioration has to occur relatively quickly. If we take the leaking gutter example, the leak must be considerable and soaking through the wall to affect internal surfaces. If the same amount of water were flowing down the face of a cavity wall, the level of urgency would be different because the damage would be much less. So, do not be over zealous when applying 'serious' or 'urgent' definitions to defects, but reserve them for defects that have a measurable impact on the building.
- A defect which presents a health and/or safety threat – This is another part of the 'urgency' test. Assessing whether a defect is a threat is a complex issue, and is considered in more detail on page 59.

Further investigations

The definition of condition rating 3 (CR3) includes the phrase: 'investigated urgently'. The practice notes [paragraph 4.5, page 17] expand on this definition and (as with the HCR before), say that a call for further investigations must result in a CR3, and state that 'Surveyors must not call for further investigations only to cover themselves against future liabilities'.

I welcome this guidance, and see it as a reaction to the HSV, where customers were often referred to numerous 'specialists' for further investigations. After paying several hundred pounds for a survey report, people are disappointed to find they still have to ask other 'specialists' to assess the property before the condition is fully explained. However, these are sentiments easier to say than to put into practice, and there will be times when further investigations are appropriate:

- where there is a suspicion that a visible defect may have affected other, concealed building elements and;
- where the extent of a visible defect is not possible to ascertain within the confines of the inspection.

In such cases, the practice notes state [paragraph 4.5, page 17] that the surveyor must '... recommend that a further investigation is undertaken and a condition rating 3 applied'. However, the practice notes also point out that surveyors should not use this approach simply because a building element is inaccessible. Instead, it could result in a rating decision based on a restricted inspection or a 'Not Inspected' (NI) rating. This is where your knowledge of building construction and pathology can be used to make carefully considered, speculative judgements. The following examples may illustrate this more clearly.

Roof space inspection

Imagine you are inspecting an older terrace property that has a slate covered, pitched roof and the bedroom ceilings below the roof space are lath and plaster. Consider the following scenarios:

Scenario One: Externally, you can see both roof slopes from the ground with your binoculars. The original slate covering is in a poor condition with a number of cracked, slipped and missing slates. Internally there is no loft access hatch so you could not inspect the roof space. However, the ceilings below the roof space are badly stained and one stained area is cracked and bulging. The stained areas register 'red' readings on your moisture meter when tested.

Action: *You have evidence that the roof covering is in poor condition and it is very likely that the staining is caused by active roof leaks. Although this is a serious and possibly urgent defect, there '… is a suspicion that a visible defect may have affected other, concealed building elements'. It would therefore be important to inspect the loft to confirm the cause and see if there are any further problems. A further investigation in such a case is justified.*

Scenario two: You can see both of the roof slopes from the ground and the slate roof covering looks to be in a satisfactory condition. There are no cracked or slipped slates, and no visible tingles. Internally, there is no loft access and the ceilings to the rooms below the roof space are in a satisfactory condition. They are level and even and without water staining.

Action: *Although the roof space has not been inspected, based on what you can see, there are no obvious signs of a problem. I could make an argument that slates of that age will often first deteriorate on the underside and all the nail fixings could have corroded and were about to fail, but that all seems too speculative. Consequently, a 'Not Inspected' rating would be appropriate.*

Scenario three: This scenario is slightly different. Assume you can see the whole of the front roof slope but only part of the rear slope because the overgrown and small rear garden prevents you from getting a good view. Where you can see the slate roof covering, it looks to be in a satisfactory condition. There are no cracked or slipped slates, and no visible tingles. Internally, you can get access to the loft where you can inspect the underside of both roof slopes including that to the rear. This reveals no cause for concern. The ceilings to the rooms below the roof space are in a satisfactory condition and are level and even and without water staining.

Action: *Most of the roof covering element can be inspected, apart from the outside of the rear roof slope, and although the slates to the rear could be in a different condition, the internal inspection should provide reassurance. The ceilings are not stained and, assuming there are no other issues, this seems to be a case where sensible speculation can be used to arrive at a clear decision. A condition rating can be allocated but it would be important to emphasise that your judgement was based on a restricted inspection.*

Scenario three shows the relationship between 'Further Investigations' and applying a condition rating with a 'restricted inspection'. Although these scenarios relate to the roof coverings, the same approach could be applied to most other elements. For example, consider suspended timber ground floors:

- Assume the floor surface is covered with a fitted carpet. The inside faces of the external walls and associated skirtings are damp, the floor vibrates excessively after a 'heel drop' test, there is no under-floor ventilation and the external ground levels match that of the internal floor surface. An obvious example, but clearly a case for 'Further Investigation' because of the likelihood of concealed defects.
- Imagine a similarly constructed floor but in the opposite condition. There is no dampness, the floor does not vibrate, there are enough under-floor air bricks and the external ground levels are at least 150mm below the floor level. Although there could be a rogue outbreak of dry rot lurking, all the signs are positive. A 'Not Inspected' would be appropriate here.
- Assume now that conditions are the same as the latter case, except that the carpet can be lifted from one corner and a loose floorboard removed. Using a torch and a hand-held mirror, 50 per cent of the under floor void can be seen. This reveals a relatively clear under-floor area with the joists bedded on the damp-proof course (DPC) and properly supported by 'honeycombed' sleeper walls. There could be concealed defects, but as all the indicators are satisfactory, a condition rating can be applied and the restriction on the inspection noted in the report.

Building services – the special status of further investigations, 'not inspected' and condition rating

Many residential practitioners offer little or no comment on the condition of the service systems in their reports. Instead, they refer their clients to specialists, almost as a matter of course. To serve our customers better, I think we have to reverse this trend and engage with these elements. However, there is a balance to be struck. Surveyors are not specialists and if not installed and maintained properly, some service installations can kill. Therefore, judgements in this area have to be carefully measured.

Standard advice

Like the HCR before it, the HBR includes mandatory text written by the Department of Communities and

Local Government (CLG) in the text box of element G1 (Electricity) and G2 (Gas and Oil). **This cannot be omitted or changed** and is shown in Table 4.2. As a consequence, all condition rating decisions and reports will have to conform to this standard advice. If not, we will leave ourselves vulnerable to challenge.

In my view, the inclusion of this standard advice in the HBR represents a significant change in how services are assessed and reported. This has proved a controversial issue in my training seminars and I will try to present the flavour of these discussions here.

A changing regulatory environment

Whether we like it or not, we operate in an increasingly regulatory environment. For all the following examples, a registered 'competent person' must do the work, or Building Regulations approval must be obtained from the local council:

- installing a new solid fuel, gas or oil fire in an existing chimney breast;
- the provision of an LPG or oil storage tank;
- the replacement of a defective hot water cylinder;
- the replacement of a gas or oil boiler;
- fitting a new electrical circuit or providing a new light in a shower room.

Other recent changes also have an impact:

- The Institution of Electrical Engineers (IEE) has recently issued the 17th Edition of the *Wiring Regulations*, which includes enhanced safety features especially in bath and shower rooms.
- CAPITA, a private sector company, has taken over the gas registration scheme from CORGI. Called the 'Gas Safe Register', this organisation also has a brief to review and reform the gas-related competent persons scheme (see www.gassaferegister.co.uk).
- The Water Regulations 1999 gave utility companies wide-ranging powers to require upgrading work to prevent contamination of water supplies.
- In 2011, the responsibility for private sewers is likely to pass from home owners to local water companies, and the Environment Agency are getting strict over the standard of outflow from private sewerage treatment plants.

Faced with these mind-boggling changes, it is not surprising that surveyors can be unaware of many requirements. As generalists who give broad advice on the suitability of a dwelling for purchase, chartered surveyors cannot be expected to have a deep understanding of these issues. However, neither can they use ignorance as a defence. This information is in the public domain and surveyors should at least be aware of it.

Increasing concern over safety issues

Some blame the growth of the 'nanny state', but the public are increasingly concerned with the safety of their home environment. This is discussed in more detail on page 59; suffice to say here that every time a gas explosion destroys a house, or a person suffocates because of a carbon monoxide leak, or someone is killed by an electrical fault, the story is all over the press. A surveyor's mistake that leads to financial loss is one thing; failure to reflect safety advice that is in the public domain is quite another. Consequently, we have to be cautious and err on the side of safety.

Annual servicing and other safety checks

For years, home owners have been advised to have their gas and oil appliances serviced annually, and their electrical systems checked regularly – but how many actually do? Even if a home owner has their boiler serviced it is rare that they include the gas fire on the same agreement as well. At training seminars, I always ask for a show of hands by those who have followed the IEE's recommendations with their own electrical systems. As you might imagine, it is rare that more than a few hands are raised.

Table 4.2: Mandatory text printed in G1 and G2

G1: Electricity

Safety warning: *You should have your electrical installations inspected and tested regularly, to protect your home from damage and to avoid putting your safety at risk. Guidance published by the Institution of Electrical Engineers recommends that electrical installations should be inspected and tested at least every 10 years and when the occupiers of the property change. All electrical work carried out after 1 January 2005 should be recorded on an Electrical Installation Certificate.*

G2: Gas/Oil

Safety Warning: *All gas and oil appliances and equipment should regularly be inspected, tested, maintained and serviced by a registered 'competent person' and in line with the manufacturer's instructions. This is important to make sure that the equipment is working correctly, to limit the risk of fire and carbon monoxide poisoning, and to prevent carbon dioxide and other greenhouse gases from leaking into the air. For more advice contact the Gas Safe Register for gas installations, and OFTEC for oil installations.*

Impact of the Home Information Pack

As already discussed, the Property Information Questionnaire (PIQ) is one of the mandatory documents in the HIP. One of the most important questions asks whether the gas appliances and electrical systems have been recently checked. I accept that owners and their agents can dodge the issue by answering 'don't know', but at least it puts the issue on their agenda. They can hardly be surprised if a surveyor asks similar questions later in the process.

A change in culture

Taken together, the influences discussed above have made the safety of service installations more important than ever. A purchaser will want to be reassured that their heating system and electrical system will not only work properly but will work safely. Responding to this requirement, surveyors must ask for evidence that the systems have been properly installed, maintained and altered, as well as visually assessing the services. If the information is not forthcoming, then further investigations must be recommended.

Until this practice becomes established and understood, it is likely to be controversial. However, as the standard advice for gas, oil and electric testing is printed in every HBR, I cannot see that we have any other choice. A number of building service case studies have been included in chapter 5 to illustrate how a balanced approach can be developed.

5. Developing a Consistent Approach – the Condition Rating Protocol

The previous chapter introduced the principles that underpin the condition rating process. However, you may have to make up to 40 individual condition rating decisions during an HBR. In order to do this quickly, accurately, and consistently, an easy-to-use methodology is required. Based on the work carried out during the development of the HCR, I have produced a 'Condition Rating Protocol' to suit the new HBR – shown below. This chapter begins by describing the Protocol; and then demonstrates how this might be applied, using a number of illustrated case studies.

Building element: _____

Stage One

Is the building element in a satisfactory condition?

☐ **Yes:** There are no indications of present or suspected defects that require the undertaking of a specific repair. It requires normal maintenance only.

☐ **No:** The building element needs to be repaired, replaced or investigated.

Stage Two

	No	Yes
Question A *Spoil the function* Does the problem impair the intended function of the building element?	☐	☐
Question B *Failure in other elements* Has the defect caused structural failure or serious defects in other building elements?	☐	☐
Question C *Structural integrity* Has the defect compromised the structural integrity of the element?	☐	☐
Question D *Health and safety* Does the defect seriously and directly threaten the health and safety of the building users?	☐	☐
Question E *Urgent repair* Are urgent repairs or replacements needed now? 'Now' means the repair or replacement cannot be left until the next anticipated period of regular maintenance without it causing structural failure or leading to a 'serious' defect in another element of the building.	☐	☐
Question F *Further investigation* Does the problem require further investigation?	☐	☐

THE CONDITION RATING PROTOCOL

Using the protocol I suggest makes deciding on the correct condition rating a straightforward process: begin with the inspection and assessment of the building element as normal; then apply the two-stage process (as illustrated):

Stage One: Is the building element in a satisfactory condition?

If there are no defects or deficiencies that require repair or replacement, a condition rating 1 (CR1) will be appropriate and further consideration will not be required. Conversely, if something more than 'normal maintenance' is needed then you should move on to Stage Two.

Stage Two: The purpose of this second stage is to decide whether a condition rating 2 or 3 (CR2 or CR3) is appropriate.

The six questions I suggest you address, to help you identify the critical features of the defect, are developed from the definitions described in the practice notes. Simply apply each question to the issue you have identified, then tick the 'yes' or 'no' box as appropriate. Once you have responded to each question, you should look at the results as a whole. For example, if you have ticked all the 'no' boxes the defect is unlikely to be serious or urgent and so CR2 would be appropriate; if you have ticked one or more 'yes' boxes, CR3 will generally be suitable.

This is not a precise science that automatically leads to a 'right answer'. Instead, I see it as a decision-making tool that gets you in the right ball-park. Although this method encourages an objective approach, there is enough flexibility for you to apply your own professional and judgment discretion in marginal cases.

BROADER ISSUES THAT AFFECT CONDITION RATING DECISIONS

Before reviewing the condition rating case studies, it will be useful to look at some other condition rating principles identified in the practice notes:

- **Refer to condition only, and do not reflect purely cosmetic issues that have no effect on longevity or performance** – When deciding on the condition rating, it is important to consider only the physical state of the element. To protect the validity of the HBR as a product, surveyors all over England and Wales need to make broadly similar rating decisions. Accounting for other issues will inevitably skew the rating. For example, imagine a ten-year-old felt-covered flat roof over a lounge extension that is not leaking. Such constructions do have a short life, but this should not affect the rating decision, and in this case it is likely to be CR1. The future implications of flat roofs should be discussed in the report, rather than having an impact on the rating.
- **Reflect any detrimental effect on surrounding building elements** – This is a restatement of one of the main criteria for CR3.
- **Reflect performance** – How well will the building element keep performing its function?
- **Compare 'like with like'** – For example, do not compare the life expectancy of a flat felted roof with a pitched tiled one. This is an important principle because there is no universal 'ideal' condition. Instead, you need to compare the building element with others of a similar type and age.
- **Assume that normal regular maintenance will be undertaken in future** – You must always assume that the person in charge of the property is a 'reasonable' owner, even if you suspect they may not be!
- **Be consistent** – Try to adopt the same approach, property-to-property, area-to-area.
- **Follow generally accepted building practices** – Do not bring your own subjective opinions to the assessment. I have found that many practitioners have their own highly individual construction preferences that often fly in the face of current practice. Make sure you are familiar with current research and construction practice and apply those standards, not your own.
- **Be reasonable; perfection is not the norm** – This is self-explanatory and you should not set unrealistically high standards.
- **Do not reflect differences in product quality unless performance and life expectancy are seriously compromised** – This is especially true with DIY elements within a property. Although you might find some property improvements extremely distasteful, do not let them affect your assessment, unless they really do affect the longevity of the element (such as a flat roof that has been covered with a nailed felt covering, or bare and untreated timber that has been used externally).
- **Disregard individual taste or fashion** – Bright orange walls may give you a headache, but it will not affect the condition rating.
- **Take serious health and safety issues into account** – See the later section in this chapter for further discussion on this aspect.

CONDITION RATING CASE STUDIES

To illustrate how these different techniques can be used, I have assembled a number of illustrated examples. These are typical examples of their type and will help you to identify the differences between the condition ratings. For each example, I have included a number of photographs, together with a description of the condition.

Condition rating of roof coverings (E2: Roof coverings)

Roof covering A

(a)

(b)

This is a clay-tiled 'pitched and hipped' roof of a house built during the inter-war period. Points to note:

- The covering is supported by a traditional timber 'cut' roof structure, which is adequate for its purpose.
- There are no missing or cracked tiles. The roof slopes are even and satisfactorily laid. The ridge and hip tiles have recently been rebedded (although not too neatly).
- The covering has no roofing felt beneath it and the tiles are back-pointed (known as 'torching' in some parts of the country) with no roof leaks. The nibs on the back of the tiles are in satisfactory condition and only a few tiles have been replaced in the past.

(c)
Figure 5.1: (a) a general view of the front of the house; (b) a view of the rear of the roof; (c) a closer view of the roof

Roof covering A – feedback

This clay tiled roof is in satisfactory condition. The tiles look well laid and the slopes are even. There are only one or two tiles that are cracked or partially missing but these have not resulted in a problem at the time of inspection. In my view, this comes within the definition of 'normal maintenance'. There is no roofing felt beneath the tiles which makes it vulnerable to wind blown rain and snow but the back pointing or torching will offer some protection from water penetration. The repointing and/or rebedding of the ridge and hip tiles suggest recent maintenance work has been carried out. The absence of roof leaks to the rooms below completes the picture and based on these photographs, I would rate this roof covering CR1. The questions in Stage Two have not been applied because they are not relevant as this is clearly CR1.

Stage One	
Is the building element in a satisfactory condition?	
✓ **Yes:** There are no indications of present or suspected defects that require the undertaking of a specific repair. It requires normal maintenance only.	☐ **No:** The building element needs to be repaired, replaced or investigated.

Roof covering B

(a)

(b)

(d)

(c)

Figure 5.2: (a) a general view of this rural cottage; (b) a closer view of the roof; (c) roof from a different angle; (d) a view of the underside of the roof covering, taken from within the roof space

This is a heavy stone slate (probably sandstone) roof covering to an isolated rural terraced cottage built approximately 200 years ago. Points to note:

- The covering is supported by the original timber 'cut' roof structure that is adequate for its purpose although some parts are slightly undulating.
- Most of the stone slates seem to have been well graded and laid in appropriate courses with no sudden changes in thickness or big gaps between adjacent courses. The edges of a number of slates are uneven but not unusual for slates of this age.
- There are no missing or cracked stone slates and the internal inspection revealed the covering has a relatively modern roofing felt beneath it and this suggests it has been recovered recently.
- There are no roof leaks to the rooms below.

Roof covering B – feedback

This older property is covered by heavy stone slates which are common to this particular region. They are relatively thick, and have to be selected and fixed carefully to ensure the gap between adjacent slates is kept to a minimum. The largest slates are laid towards the eaves with the smallest towards the ridge. The weight of this type of roof covering can cause deformation of the supporting timbers and, although a certain amount of unevenness is normal, if it becomes too pronounced, the structure may have to be repaired. Older coverings are often fixed with oak or other wooden pegs. These can shrink and rot, causing the individual slates to slip down the roof.

In this example, the relatively modern roofing felt beneath the covering indicates that this roof covering has been completely re-laid in the past five to ten years. Second-hand stone slates have been used and these seem to have been appropriately selected and laid. Lead flashings have been used at the junction with neighbouring roofs and chimneys. Although some of the detailing appears to be crude there are no leaks to the rooms or spaces below. Although this type of roof covering may have higher maintenance costs than modern equivalents, it is in a satisfactory condition for its type. Therefore, in my opinion a repair is not necessary and this should be marked CR1.

This example highlights two important principles for condition rating:

- You must always compare like with like. If you judge the stone slate covering against that of clay tiles, I am sure most surveyors would give the stone slates a higher rating.
- You must be familiar with the particular construction methods and materials so that you can come to a balanced decision. If not, then your assessment is likely to be unsafe. This will always be a matter of sensible judgement. For example, although I have assessed a number of stone slate roofs for a variety of purposes I would not call myself an expert. Because of this, I would take a greater number of photos and make more thorough notes than I normally would and, once back at the office, I would check the standard texts and other information to make sure I was assessing the covering correctly. On the other hand, put me in front of a thatched roof and I would be lost! True; I could adopt a similar approach, but as I have never assessed thatch before, I would have to declare it outside of my area of competence.

Based on this discussion I would rate the slate roof CR1. As with Roof A, I did not apply the questions in Stage Two of my protocol because this is clearly CR1.

Stage One
Is the building element in a satisfactory condition?
✓ **Yes:** There are no indications of present or suspected defects that require the undertaking of a specific repair. It requires normal maintenance only. ☐ **No:** The building element needs to be repaired, replaced or investigated.

Roof covering C

This is an original slate roof covering to a detached house built in 1913 (shown overleaf). Points to note:

- The covering is supported by the original purlin and rafter roof structure. The property has an original room-in-the-roof, which has a dormer window with a felt-covered flat roof (the felted eaves detail can be seen from the ground with binoculars).
- There is no roofing felt beneath the slates and back pointing (torching) is in poor condition and dropping away in a number of areas.
- There are a small number of cracked, clipped or replaced slates to the roof slopes. To the rear slope, there is a missing slate and a roof leak has stained the bedroom ceiling below (see Figure 5.3(d)). Assume that high moisture meter readings are obtained in these stained areas.

A Surveyor's Guide to the RICS HomeBuyer Service

Figure 5.3: (a) a general view of the detached house; (b) a view from the window of the dormer looking towards the gutter; (c) the underside of the roof slope; (d) The ceiling beneath the slate shown in (b)

Roof covering C – feedback

Professionally, I am more comfortable with slate roofing as a construction type, and this one is almost certainly original to the house. Although it is not the worst I have seen, it does seem to be approaching the end of its useful life. There is evidence of previous repair, and it has one active roof leak that has stained the ceiling of the room below. Under Stage One of the protocol, there is clearly a defect that needs to be repaired (see Figure 5.3(b)) so the questions under Stage Two should be applied (as shown opposite). My explanation for these decisions is presented below.

5. Developing a Consistent Approach – the Condition Rating Protocol

Stage One
Is the building element in a satisfactory condition?

☐ **Yes:** There are no indications of present or suspected defects that require the undertaking of a specific repair. It requires normal maintenance only.   ✓ **No:** The building element needs to be repaired, replaced or investigated.

Stage Two	No	Yes
Question A *Spoil the function* Does the problem impair the intended function of the building element?	☐	✓
Question B *Failure in other elements* Has the defect caused structural failure or serious defects in other building elements?	☐	✓
Question C *Structural integrity* Has the defect compromised the structural integrity of the element?	✓	☐
Question D *Health and safety* Does the defect seriously and directly threaten the health and safety of the building users?	☐	✓
Question E *Urgent repair* Are urgent repairs or replacements needed now? 'Now' means the repair or replacement cannot be left until the next anticipated period of regular maintenance without it causing structural failure or leading to a 'serious' defect in another element of the building.	☐	✓
Question F *Further investigation* Does the problem require further investigation?	✓	☐

Question A – The primary function of a roof is to keep the water out. This defect has allowed water to enter the building and affect the ceiling before. Judging by the size of the stain, this is a minor leak, but it has impaired the function all the same.

Question B – Following on from the above, a defect in the roof covering element is affecting the ceiling element, although I would not describe it as a serious problem yet. The ceiling is stained but it is not collapsing. In my view, this is a marginal decision but because it is a 'yes' or 'no' response, I have answered 'yes'. Most home owners would be concerned by an active roof leak and would find it hard to understand why their surveyor did not clearly flag it up.

Question C – Looking at the roof covering as a whole, although one missing slate has caused the problems below, it does not threaten the structural integrity of the roof as a whole. Surveyors will often point out that strong winds can easily convert a missing slate to one cavernous gaping hole but I think this unlikely; it sounds a little too speculative. Therefore, I answered 'no' to this question.

Question D – A loose slate can be a danger to the users of the building, especially if it lands on their head! This flippant comment underlines a decision I often find difficult; what is 'a health and/or safety threat'? Although I discuss this in some detail later in this chapter, I still have to make a decision here. Figure 5.3(b) clearly shows a detached slate sitting in the gutter; and directly below is the rear entrance door and the back yard. Although the rear door is unlikely to be used as much as the front, I think the likelihood of someone being in the vicinity when this slate comes crashing down is high enough to affect the rating (but only just). Therefore I have answered 'yes' to this question. Please see page 59 for more discussion of this matter.

Question E – Water leaking through roofs onto the ceiling below can cause a range of problems, from unsightly staining, wood rot in the ceiling joists and timber laths, to complete collapse of the ceiling. In my opinion, this minor leak has not yet caused problems of this magnitude, but what if there is a particularly heavy downpour? Additionally, I previously identified (albeit reluctantly) the loose slate as a possible safety risk and there is a body of thought that considers a safety risk has to be resolved urgently. Therefore, the repair cannot be left until the next planned maintenance work, and it needs to be done now. Therefore I ticked 'yes' to this response.

Question F – Where you can see all roof slopes from the ground and inspect the underside of the roof covering, you should be able to come to a clear view about the condition of the element and so this does not require further investigation. The new owner may want to invite a roofer to provide an estimate of the repair but that is a different matter completely.

Looking at the profile of the responses, the four 'yes' answers justify allocating a CR3 to the element. However, because my responses under A, B and D were marginal I feel it was only just in the CR3 category.

Condition rating the flat roof over the dormer window

Although this case study is focused on the condition of the pitched roof, Figure 5.3(a) shows that there is a second roof type: the flat roof over the dormer window. There are two points to note here:

- It is a different type of roof and so should be classified separately from the pitched slate roof and allocated a rating of its own.

- The roof covering is not overlooked from a higher window and only the edge of the felt roof covering can be seen from the ground with binoculars.

Assuming that there are no roof leaks showing to the ceiling beneath, the advice described on page 41 can be applied and, in this case, I concluded:

- There is no evidence of roof leaks and the felt eaves detail gave no cause for concern. Consequently, I saw no reason to call for a further investigation.

- Although seeing the felt around the edge is helpful, it does not provide enough information on which to base a 'conditional' rating.

- Therefore, the appropriate decision would be to allocate a 'Not Inspected' (NI) to this roof and make sure that the reasons are clearly given in the appropriate section.

However, that is not the end of it. As described on page 84, you can still offer some general and broad advice on the roof.

Condition rating of heating systems (G4: Heating)

The difficult job of condition rating building services systems has already been discussed in chapter 4. In this case study, I want to show how the protocol can be used to rate different gas central heating systems. To keep the presentation of the case study information manageable, I have focused on the boilers. In reality, the controls, pipework, radiators and ancillary equipment would be included in the assessment.

Gas boiler A

(a)

(b)

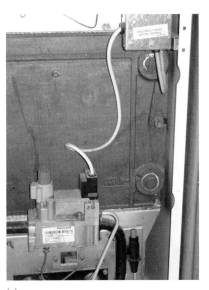
(c)

(d)

Figure 5.4: (a) gas boiler A; (b) with the front panel removed; (c) a close-up of boiler A; (d) the flue that serves boiler A

This is a floor-standing gas boiler, installed in the kitchen of a terraced house built in approximately 1920. At the time of inspection, the property was empty and the keys were obtained from the agents who knew very little about the property. The Property Information Questionnaire (PIQ) in the Home Information Pack was not available and there was no documentation at the property. The front of the boiler casing is partially detached and can be removed to allow an inspection of the internal components. There are no visual indications of a problem. The balanced flue from the boiler discharges directly through the wall into the rear yard. The bottom of the flue casing is approximately 500mm off ground level.

Gas boiler A – feedback

This is an older boiler – at least ten years old and probably closer to 20. Most maintenance organisations plan for boiler replacements every 10–15 years, so this one is doing well, but will be close to the end of its economic life. Although there are no visual signs of a problem, because the property is empty, I do not know whether it works. Points to note:

- The lack of documentation is not surprising but poses a significant problem. Without any evidence that the boiler has been installed and serviced regularly, it is impossible to make a judgement about its safety.
- Although the photographs show the front of the boiler casing (it just fell away in my hands), it should only be opened if designed for that purpose. For example, if the front panel is hinged and secured with magnetic catches, then it should be opened. However, if the panel has to be 'uncoupled' with difficulty then it should not be removed because this goes beyond the extent of the normal

inspection and could create a health and safety hazard.
- The external balanced flue is relatively close to the ground and at this level the hot surfaces could easily be touched accidentally and cause injury. Therefore, it should be protected by a stout wire cage.

Taking these points into account, the lack of maintenance history and the position of the unprotected flue means that some work or further investigation is required. Therefore I proceeded directly to Stage Two and my responses are explained below.

Stage Two	No	Yes
Question A *Spoil the function* Does the problem impair the intended function of the building element?	☐	✓
Question B *Failure in other elements* Has the defect caused structural failure or serious defects in other building elements?	✓	☐
Question C *Structural integrity* Has the defect compromised the structural integrity of the element?	✓	☐
Question D *Health and safety* Does the problem seriously threaten the health and safety of the building users?	☐	✓
Question E *Urgent repair* Can the repair or replacement be left until the next anticipated period of regular maintenance without it causing structural failure or leading to a 'serious' defect in another element of the building?	✓	☐
Question F *Further investigation* Does the problem require further investigation?	☐	✓

Question A – The primary function of a boiler is to heat the dwelling. This boiler is switched off, so it is not possible to say whether the boiler works or not. However, this is a 'point in time' inspection and because the boiler is not working now, it could be argued that it does not fulfil its primary purpose. Consequently, I ticked the 'yes' response.

Question B – This question is not really relevant to a boiler. If I wanted to be obtuse, I could hypothesise that the boiler would affect other building elements especially if it exploded. As this is unlikely, I would classify this view as 'too speculative'.

Question C – As above, this question is not relevant for a boiler.

Question D – Based on the advice from organisations such as CORGI, the Gas Safe Register and the HBR practice notes [Section J3, page 65]; gas appliances that have no safety record present a danger to the occupants. Therefore I have ticked the 'yes' box.

Question E – I have ticked the 'no' box here. I know this contradicts what I said earlier about the loose slate, but in this example I ticked 'no' because I do not know whether there is a problem with the boiler –it is turned off and there is no maintenance record. Instead, further investigations are required to establish the true condition before a decision can be taken on whether urgent remedial works are required.

Question F – Establishing the condition of a gas boiler requires specialist skills and knowledge and is beyond the competence and professional role of the surveyor. Therefore, a tick in the 'yes' box is appropriate.

Looking at this set of responses, CR3 is appropriate and would come as little surprise to the seller, buyer or even the agent!

Gas boiler B

(a)

(b)

(c)

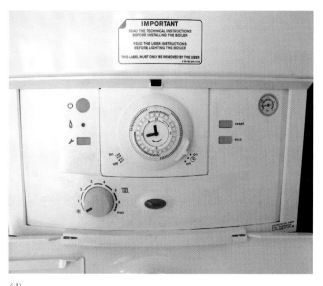
(d)

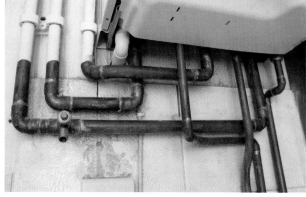
(e)

Figure 5.5: (a) gas boiler B; (b) external view showing the house of boiler B (on the right), the passageway shared with the neighbouring property, and the white tubular balanced flue to boiler B; (c) detail of the external pipework; (d) gas boiler B with its hinged panel open; (e) the pipework below boiler B

Gas boiler B – feedback

This is a modern condensing boiler, identified by the plastic condensate pipe that emerges from the underside of the boiler and discharges externally. Points to note:

- Like the first example, the agent denies all knowledge of the property but at least the PIQ is available. However, this is of little use because the 'don't know' option was checked. This enables the owner to meet their legal requirements under the HIP legislation but offers little useful information for potential buyers or their advisers.
- Although this is a modern boiler (almost certainly less than five years old), there is still no proof that it has been properly installed and maintained. Visually, I could not see any major problems, but I am not a registered gas engineer. Faults that result in gas escaping or a carbon monoxide leak are not usually easy to see and so it is unlikely that surveyors will spot them.

Many surveyors are concerned that other players in the property sector – sellers, agents, financial advisers and lenders (to mention but a few) – see us a 'deal breakers'.

This wall-mounted gas boiler is also located in the kitchen of an end-of-terrace property built in the 1930s. The dwelling is empty and like the previous property above, the key was obtained from the agent who claimed that their colleague who knew the details of the sale was on holiday. The Property Information Questionnaire is available and the 'don't know' box is ticked in response to the question 'When was your central heating system last serviced?' There is no documentation at the property. The flue and other associated pipework goes through the external wall and discharges into the shared passageway along the side of the house (see Figure 5.5(c)).

The parties whose interests are served by the purchase going through as quickly as possible will often positively discourage buyers from appointing a surveyor to look at the property. In this case, if the surveyor allocates CR3 and calls for further investigation, even though the property has been marketed with '… a recently installed, energy efficient gas central heating system…', I can understand the 'disappointment' that the parties to the sale may feel. But what is the alternative? Not to emphasise the potential danger and just hope it does not cause a problem when our clients move in? Personally, I would rather face the wrath of any angry agent than have to give evidence to a Coroner's court.

Since 1998 and the subsequent changes in Part L of the Building Regulations, condensing boilers have become the norm and both contractors and surveyors are more familiar with their particular features. Although this installation looks generally satisfactory, I am concerned about two features:

- The condensate discharge pipe should be taken to a formal drainage point or a properly constructed soakaway. In this case, although the pipe is insulated externally to stop it freezing (not often done) it appears to discharge over the ground, During the inspection, I could find no evidence of a drainage connection or a soakaway.
- The position of balanced flues is governed by the Building Regulations, and the following factors may affect the flue position in this case:
    - being opposite another balanced flue;
    - facing over a neighbouring property and;
    - being located within a narrow passageway.

Apart from spotting these potential problems, I cannot come to a judgement because it is beyond my knowledge. If I had sight of documents that showed a registered gas engineer had serviced this system within the last 12 months, then I would feel reassured. In the absence of this, these features simply deepen my concern about the system as a whole.

In a similar way to the previous example, I have included stage two of the protocol together with a brief explanation.

Stage Two	No	Yes
Question A *Spoil the function* Does the problem impair the intended function of the building element?	☐	✓
Question B *Failure in other elements* Has the defect caused structural failure or serious defects in other building elements?	✓	☐
Question C *Structural integrity* Has the defect compromised the structural integrity of the element?	✓	☐
Question D *Health and safety* Does the problem seriously threaten the health and safety of the building users?	☐	✓
Question E *Urgent repair* Can the repair or replacement be left until the next anticipated period of regular maintenance without it causing structural failure or leading to a 'serious' defect in another element of the building?	✓	☐
Question F *Further investigation* Does the problem require further investigation?	☐	✓

Questions A, B and C – I have ticked the 'yes' response for the same reasons as boiler A.

Question D – Although this is a modern boiler, the lack of a safety record combined with doubts over the condensate drain and flue position raises questions over the installation's safety. Therefore I have ticked the 'yes' box.

Question E – Further investigations are required to establish the safety of the system before a decision can be taken on whether urgent remedial works are required, so I have ticked the 'no' box.

Question F – For the reasons previously given, the 'yes' box has been ticked.

Looking at this set of responses, CR3 is appropriate and – in this case – that would come as a surprise to the seller and agent and possibly even the buyer too!

Gas boiler C

(a)

(b)

(c)

Figure 5.6: (a) boiler C; (b) a closer view of the pipework; (c) a view of the flue from boiler C close to the ridge of the house; (d) the service documents

(d)

This wall-mounted gas boiler is in a cupboard on the first floor landing of an end-of-terrace house, built in 1980. To the left hand side of the boiler is a false wall (or boxing) that conceals the soil and vent pipe into which the sanitary fittings from the bathroom discharge. The flue from the boiler extends vertically through the roof space and passes through the roof covering near to the ridge. Figure 5.6(c) shows the flue terminal and the nearby soil and vent pipe. The dwelling is occupied by the current owner who states that the boiler was installed by a registered gas engineer four years ago and has been serviced every year under a British Gas service agreement. The owner produced a properly completed 'Benchmark' log book and several annual service reports from British Gas, the last one of which was dated two months previously.

Gas boiler C – feedback

This is very similar to boiler B: it is a high-efficiency wall-mounted condensing gas boiler with a balanced flue. Assume that the condensate drain discharges into the internal soil and vent pipe and the vertical balanced flue discharges at roof level. If you have a broad knowledge of modern boilers then there is little to worry you here. However, a number of surveyors will be looking at the photos with some serious concerns. Here are my guesses as to what these could be:

- the boiler has a vertical open flue, and needs a lot of combustion ventilation;
- the cupboard does not have any air grilles;
- the condensate drain has to be taken to an external gulley.

Although these concerns are loosely based on fact, *they are very much out of date*. For example, it is a room-sealed boiler. This means that it has a 'balanced flue' which serves two purposes: to draw in air for the combustion process *and* to expel combustion gases. With fan assistance, these flues can extend for many metres enabling the boilers to be sited in a greater range of locations.

Condensing boilers operate at lower temperatures so most can be located in a cupboard that has no ventilation. In fact, many manufacturers have designed their boilers specifically to be hidden away. Finally, although the pressure relief pipe from a sealed system must discharge externally, the condensate drain can connect to a range of internal plumbing systems.

In any event, unlike the previous case studies, this property is occupied and the owner has sensibly retained proof that the system has been not only been properly installed but is regularly serviced. In this case, your role is to satisfy yourself that the installation/service documentation is authentic, appropriate and current. Here are some tips on this process:

- Do not phone up certification schemes to check an engineer's documentation. Instead, make a judgement based on the information provided. Original versions of British Gas reports will be more acceptable than scruffily written notes on plain paper from dubious 'engineers'.
- Make sure that the documentation covers *all* the appliances in the dwelling. Many home owners understand the need to service their boilers, but few include their gas fire in the arrangement yet this can be just as dangerous.
- Gas (and oil and solid fuel) appliances should be serviced annually. Check that the agreement is not past its 'sell by' date.
- If the documentation is acceptable, make sure that the system has not been altered, damaged or affected by new faults since it was last seen. Even when you are satisfied the appliances are well maintained, your job has not quite finished. What if someone has altered a gas pipe or damaged the flue since the engineer last visited? You still have a duty to check.

In the case of boiler C:

- the boiler has been properly installed;
- all the appropriate documentation is in place; and
- I could see no evidence that the system has been either altered or damaged.

Therefore I would mark this CR1. Surprised? I hope not; and at least it will keep the agent happy.

Condition ratings and building services – a case of extremes

The gas boiler case studies have raised a number of issues about the condition rating process for building service systems in residential dwellings – in particular those that involve gas, oil, solid fuel and electricity. In other words those systems which, if they go wrong, can kill.

I do not want to sound over-dramatic, but a relatively minor fault in these systems may push a system into a CR3. In such cases, the defect does not have to be serious; it can be relatively minor and still create a hazard. The classic case is the broken electrical power point that can be dangerous and urgent, but relatively easy and cheap to repair.

Surveyors do not have the specialist knowledge to take various decisions. With gas, oil, solid fuel and electricity we have to play safe. Consequently, these systems are either satisfactory (CR1) or a potential hazard (CR3). I find it difficult to imagine an example where I would allocate a CR2 to an electrical system. It would be like saying 'I've seen a problem with your electrical system but it is not serious, get it repaired soon'. This sounds unwise, especially when the fault can lead to a safety hazard.

Consequently, it is my firm view that G1 (electricity), G2 (gas and oil), G4 (heating) and in some circumstances G5 (water heating) should only be allocated CR1 or CR3. Here are the situations where I think these ratings are appropriate:

CR1 is appropriate when:

- there are no visual indicators of a defect and the system appears to meet current standards (as far as a surveyor can judge);
- there is clear documentary evidence that the system has been properly installed, maintained and recently tested; and

- there is no evidence of DIY alterations or damage since the last test or inspection was carried out.

Under these circumstances, further investigations would not be required.

CR3 is appropriate when:
- there are visual indicators of defects that pose a serious health and safety risk and the system does not appear to be up to current standards (as far as a surveyor can judge);
- there is no documentary evidence that the system has been properly installed, maintained or recently tested; and/or
- there is evidence of DIY alterations and/or damage since the last test was carried out.

Under these circumstances, a further investigation would be appropriate.

Other services

This 'one or three' rule is appropriate for the immediately dangerous services. However, the plumbing, water, and drainage services are not so critical, are generally less specialised, and more within the competence areas of surveyors. In these cases, CR2 may be appropriate. For example, the following defects might be CR2 problems:

- cracked benching in an inspection chamber that has not become dislodged or is causing a leak;
- a soil and vent pipe that is cracked and partially rusted, but not yet leaking; or
- a water pipe work that has not been properly fixed, but is not leaking.

In each case, a repair is required that is not urgent or serious and unlikely to cause a health or safety hazard.

HEALTH AND SAFETY ISSUES AND THE CONDITION RATING SYSTEM

The practice notes make it clear that the issue of health and safety of the occupants has moved up the scale of importance compared to previous survey types. For example, the HSV practice notes (2005) mention the issue of safety when describing 'urgent works' giving just one example '… e.g. a visible broken power point'; while other parts of the document mention asbestos and '… other hazardous materials'. So surveyors have generally focused on 'life and limb' safety issues (such as gas leaks, dangerous electrics and unstable walls).

Fuelled by public concerns and safety campaigns, surveyors have gradually included a wider range of hazards, culminating in the list of issues identified in the guidance notes produced for the Home Condition Report.

The practice notes for the new HomeBuyer Service 2009 mention safety issues in several places, including:

- A defect which presents a health and/or safety threat is one of the criteria for a 'serious defect' [page 10];
- 'The surveyor should … take health and safety aspects into account' [page 11];
- '… those hazards that pose a direct threat to the users of the dwelling'. [Section J3 Risks to people, page 65].

The explanatory notes for section J go on to say that the risk should be clearly identifiable and not too remote, and the following examples are given:

- asbestos;
- lack of windows for emergency escape;
- inadequate fire precautions;
- absence of safety glass;
- lead water pipes;
- lack of safety rails, steep stairs;
- gas leaks;
- dangerous electrics;
- unsafe parts of the building;
- absence of test certificates for services;
- overhead power lines (EMFs);
- serious and significant tripping hazards;
- high radon levels;
- inappropriate use of accommodation (e.g. non-conforming loft conversion, basement bedroom, etc.);
- unprotected garden ponds.

There are few surprises here and surveyors have been identifying most of these issues for some time. However, we have to face the problem of getting the balance right.

I have already mentioned earlier in this chapter that surveyors are seen as 'deal breakers'. If we now allocate a CR3 to any potential hazard, we may exasperate those who are already confounded. On the other hand, the public are concerned about a greater range of health and safety issues. They will want to know about the potential hazards in their new home, and organisations such as the Health and Safety Executive and the Royal Society for the Prevention of Accidents (RoSPA) frequently launch campaigns to raise public awareness of hazards in the home. Consequently, surveyors need a method of arriving at a balanced and consistent rating decision.

A number of health and safety risk assessment systems already exist. For example, the government has developed the Housing Health and Safety Rating System (HHSRS) for evaluating potential risks to health and safety from any defects or deficiencies identified in dwellings. The HHSRS has replaced the old 'fitness' standards used by environmental health officers. However, this is a sophisticated technique that goes beyond the scope of the HBR service. Consequently, a more practical method is required.

Although the practice notes are a practical guide to producing an HBR, the explanation of how to assess a health and safety hazard is under-developed. For example, the practice notes state that a CR3 should be considered when a 'defect ... presents a health and safety hazard'; and under section J3 Risks to people, we have '... [there might also be] issues that may have existed for a long time and cannot reasonably be changed but still may present a health and safety risk'. There is no other discussion of how we make these judgements. Consequently, I will offer a decision-making strategy, below.

An objective method for assessing health and safety hazards

In order to develop this method, I have made two main assumptions. Firstly, hazards that are caused by physical defects are assessed within the condition rating system. This is both logical and reasonable. Whatever the circumstances, a defect that could injure or kill must be identified, assessed, emphasised and reported. Secondly, deficiencies or hazards that '... have existed for a long time and cannot be reasonably changed ...' should not affect the condition rating. This allows a balanced assessment of properties that were built before any coherent building codes existed.

To explain my second assumption in more detail, consider the example of a 250-year-old rural cottage that has very steep and narrow stairs, low ceiling heights (and even lower door heads) and several changes of floor level, which would be unreasonably 'punished' if these deficiencies were condition rated. The steep stair could be in excellent physical condition; its only drawback is that people are more likely to fall down it and die. Without an objective assessment method, such a property would achieve a high proportion of CR3s, dissuading many from proceeding. However – and this is very important – the surveyor's role is to make sure that anyone who does proceed to buy this type of property fully understands that it includes a greater number of inherent hazards than an equivalent modern property and many of these simply cannot be altered.

Armed with my two assumptions, and experience with the HCR and other survey formats which prove to me that simple and consistent assessment methods work best, I propose a three-step method, set out in Table 5.1.

Regarding benchmarks: note that you only need a general awareness, because most of these benchmarks are not retrospective and many older properties will fall below these standards. The issue to judge is how far the element falls below this measure and so becomes a serious hazard. For example, for staircases, think about: the allowable pitch of the staircase; the basic relationship between the risers and treads (including tapered treads); and broad provisions regarding handrails, balusters, guarding, etc. Then compare the staircase you are assessing with the 'benchmark'. Two responses are likely: (a) the staircase is very similar to the current benchmark, and where there are differences, they do not affect safety; or (b) the staircase

Table 5.1: A three-step methodology for assessing and reporting on health and safety hazards

Step one – Identify the potential hazard	To keep this simple, use the list of risks identified in the practice notes (also shown on page 59 of this book.)	
Step two – Does the hazard pose a direct threat to the users of the dwelling?	The HBR practice notes say that a hazard has to be clear and identifiable and should not be too remote. To complete Step two, you will need to be broadly aware of the current 'benchmarks' – the relevant standards and regulations that set safety standards in modern buildings including Building Regulations, British Standards, Codes of Practice and other publications from professional or trade bodies (for example, BRE, IEE, CORGI, Gas Safe Register).	
Step three – Action	There are two courses of action:	
	If a defect has created the hazard, you should:	If the hazard is not the result of a defect, and falls below the 'benchmark', you should;
	● include the hazard in the condition rating process; ● report the issue in the element to which it is primarily associated; ● report the hazard in section J3.	● NOT include the hazard in the condition rating process; ● report the issue in the element to which it is primarily associated; ● report the hazard in section J3.

is well below the current benchmark and the deficiencies will affect the condition rating.

One of the best ways of explaining a process is to show how to apply it in practice. A number of examples are included below; and for each one I have included a description of the hazard, discussed some of the critical issues, and outlined what action I would take. In the 'action' section, I have included both the condition rating and the relevant part of the report. For more details of my approach to report writing, see page 81.

Lack of safety glazing

Figure 5.7

These fully glazed doors are in the ground floor hallway of an inter-war semi-detached house. There are no physical defects and the doors are glazed with older, patterned glass. In my view, this is clearly a safety hazard. The doors will not have been glazed with safety glass and are on the major circulation routes in the dwelling. This is a direct threat to the users. Assuming that there are no defects with the doors, it would be reported as follows:

F7 Woodwork
Internal doors   The internal doors to the ground floor are made of timber and fully glazed: CR1.   The internal doors have glazing at low level. This is a hazard for those who fall against it and it needs to be made safe immediately.

J3 Risks to people
F7: Internal doors – lack of safety glazing at low level.

Lack of windows for emergency escape

Figure 5.8

This is a uPVC replacement double-glazed window to the small bedroom of a typical three-bedroom semi-detached house. The dwelling is empty and there is no information about the replacement windows, the dates installed or any guarantees, etc. Only the top part of the window opens and the dwelling is not fitted with smoke detectors. In training seminars, this has been a controversial issue. A fire would quickly trap the occupants in the bedroom and they would not be able to escape or be easily rescued through the window. There are several considerations:

- Without the details of the installation, we do not know whether it is has building regulation approval or is covered by a FENSA certificate.
- All regulatory bodies recommend that escape windows are incorporated in new installations, but this is not a strict mandatory requirement.
- Many local authorities would accept this type of window as long as hard-wired smoke alarms were also fitted.

Therefore, this is not a defect – it is a 'deficiency'. The window is not falling out, the double-glazed unit has not failed, the top-light opens and closes, and the rain is not getting in around the frame or the sill. Therefore the hazard will not affect the condition rating but will be emphasised in the report.

Assuming that there are no defects with this or the other windows, report as follows:

E5 Windows
The double glazed windows are made of uPVC: CR1   The window opening to the front bedroom is small. In a fire, smoke can quickly trap occupants in the room and the size of this opening will prevent them escaping or others rescuing them. To reduce this hazard, you should ask an appropriately qualified person for advice.

You should ask your legal adviser to check whether this work has received Building Regulations approval or has a FENSA certificate. If the installation has neither of these, or was installed before April 2002, you should ask an appropriately qualified person to assess the quality of the installation.

J3 Risks to people

E5: Windows – Lack of escape windows to the small bedroom.

Garden ponds

Figure 5.9

This is a 'brimming pond' (i.e. a pond that is filled right to the top, so the water is level with the surrounding ground) in the rear garden of a detached house. It is approximately 650mm deep, 1.5m wide and 3m long. There are no barriers around this feature and the paving around the pond is in a satisfactory condition.

Garden ponds are part of the 'grounds' and would come under H3 of the report and are not usually condition rated. This pond looks satisfactorily constructed and maintained, but some 10–12 people are drowned in garden ponds every year, many of them young children. Additionally, organisations such as RoSPA warn the public about the hazards of garden ponds. During training seminars, many surveyors say that the risk is obvious and it is up to the purchaser to consider it. I have sympathy with this, but we are now in a regulatory environment where it is appropriate to offer consistent and comprehensive advice on all hazards. The danger of water in gardens is clearly in the public domain and so we need to account for it.

This is another example of a deficiency and so should not be included in the condition rating assessment. However, it should be emphasised in the report under H3. Consequently, assuming that the pond is in a satisfactory condition, I would report on it as follows:

H3 General

Garden pond
The garden pond in the rear garden is uncovered and is a safety hazard, especially for young children. You may want to consider making this safer.

J3 Risks to people

H3: Garden pond – uncovered and a safety hazard for users.

Asbestos lagging to flue

Figure 5.10

This photograph shows a vertical 'open' flue from a gas fire and back boiler in the lounge of a BISF (British Iron and Steel Federation) house built in the 1950s. This flue extends up through the first floor, the roof space, and discharges close to the ridge. A soft material insulates the flue in the roof space and a wire mesh holds it in position. The presence of asbestos-containing materials is potentially dangerous and many clients are not happy sharing their home with the material. However, if it is in a satisfactory condition, and will not be disturbed in the future, it can be left in place – providing you tell the client it is there.

In this particular case, adopting that approach is not an option because this type of insulating material is likely to contain either blue or brown asbestos fibres. This is the most dangerous type and if it is in a poor condition, asbestos fibres may already contaminate the roof space.

The current occupier uses the space for storage and the photo shows a water tank behind the flue. To maintain this, any operative would have to squeeze past the insulation, inevitably disturbing asbestos fibres. Consequently, the advice has to include a clear emphasis. Because the flue coating is in a poor condition, it can be incorporated within the condition rating system. Assuming that there are no other problems with the flue, report as follows:

F5 Fireplaces, chimney breasts and flues

The property has a vertical flue that serves the gas fire and back boiler in the lounge. In the roof space, a material that may contain asbestos fibres is wrapped around the flue and is in a poor condition. This is a health hazard: CR3.

Soft insulating materials that contain asbestos are a health hazard especially when in a poor and disturbed condition. You should ask an appropriately qualified person for advice. No one should go into the roof space until it is safe. Although I do not know the extent of the work, solving the problem will be costly.

J3 Risks to people

F5: Fireplaces, chimney breasts and flues – presence of asbestos in roof space.

Asbestos soffit panels

Figure 5.11

Cupped screws secure the soffit board beneath the eaves of this dwelling, which was built in 1979. This suggests a mineral board dating from a time when the use of 'asbestolux' was common. The paint finish to the soffit and adjacent timber is beginning to flake and peel, and there are small areas of wet rot to a few sills.

Unlike the previous example, this possible asbestos-based material is in a reasonable condition and not so dangerous. It is outside and, unless it is disturbed, it will not be a health and safety hazard. However, the soffit and adjacent surfaces need to be redecorated and the last thing you want is for the owner or their contractor to get the orbital sander out to start on the preparation stage. Consequently, the report needs to identify this issue and also warn that doing the work safely can be costly. Assuming that there are no other problems, report as follows:

E8: Other joinery and finishes

The external joinery includes fascias, soffits, bargeboards, and timber cladding. The decorations are flaking and peeling, and small areas of timber are rotten. The soffit board is made of a material that may contain asbestos fibres. You should not drill or disturb this without prior advice from an appropriately experienced or qualified person: CR2.

Without a protective decorative finish, timber will rot, requiring costly repairs. To prevent this, you should redecorate the external surfaces soon. The cost will be increased if the soffit contains asbestos fibres. To repair joinery safely, contractors will have to use appropriate access equipment (for example, scaffolding, hydraulic platforms) and this can increase the cost of the work.

J3 Risks to people

E8: Other joinery – presence of asbestos.

Lead pipes

Figure 5.12

This hot water cylinder is in an older property, which needs considerable upgrading. It is a replacement cylinder fitted to an older plumbing system consisting of the original lead pipes to both hot and cold water. Short lengths of copper pipes connect the cylinder to the lead pipe system; assume there are no leaks or other defects to the water system.

Lead is a health hazard, but many of the communication pipes from the utility company's mains are made of lead. Consequently, there has to be a balance to your advice. Commonly, if the lead piping finishes as the rising main enters the dwelling, most surveyors identify its presence, describe the general health risks, and recommend replacement when convenient.

Where the lead pipes form part of the internal water systems, this will cause greater maintenance problems (the pipes bend, causing the joints to leak), and this also increases the amount of lead in the water (especially true for hot water pipes). In this case, apart from short sections connected to the hot water tank, most of the plumbing system is likely to consist of lead.

The main challenge with this example (based on the information presented) is that there are no defects. It is another case of emphasising the problem through the report *only*. Assuming that there are no other problems, report as follows:

G5 Water heating

The hot water is provided by an on-peak electric immersion heater to a hot water storage cylinder in the cupboard to the rear bedroom. Some of the internal hot pipes are made of lead and this may be a safety hazard: CR1.

I found that lead pipes form part of the hot water plumbing system. Not only does this increase the risk of lead getting into the water causing a health hazard but it is also a maintenance problem. Lead pipes can distort over time resulting in damaging water leaks. You should ask an appropriately qualified person to replace the internal lead piping soon (see also G3 Water).

J3 Risks to people

G5: Water heating – lead pipes on the hot water system.

Serious and significant tripping hazards

Figure 5.13

The photograph shows a single step between the kitchen and the lounge/diner of a small, older cottage. There is no door on the opening and the lounge floor is 90mm higher than the floor in the kitchen. The flooring to both rooms is in a satisfactory condition.

Slip, trip and fall accidents are one of the most common types of accident people have in the home and can cause very serious injuries. Older people are especially vulnerable. Changes in level of less than 300mm (usually a single step) can cause many accidents, because they are not obvious and often catch people by surprise. Where they pose a problem, removing the hazard is expensive: adjusting the relative level of the floors in different rooms is also impractical.

In this case, there is no defect, so the hazard will not affect the condition rating, but it should be emphasised in the report. I think it is important to do this because occupants could be carrying hot food and drinks between the two rooms, enhancing the risk. Assuming that there are no other problems, report as follows:

F4 Floors

The ground floor is a mixture of: solid ground floors finished with ceramic tiles, and timber floors with varnished floor boarding: CR1.

The step between the kitchen and the lounge is a tripping hazard and this could cause accidents.

J3 Risks to people

F4: Floors – tripping hazard.

Unsafe parts of the building

Figure 5.14

For details, please refer to the details of the roof case study, described on page 50, where this example has also been used. Earlier in this book, I had allocated a CR3 to this defect because the loose slate is above the rear entrance door to the property. If the slate was over a little-used side garden, you might consider choosing a CR2 because the likelihood of an accident would be lower. This is not a risk assessment process (well, not a very good one anyway) and you will always need to take a professional judgement based on what you find at the property. Caution: even though the slate could be over a quiet garden, look out for a public path or the neighbour's patio just over the fence! Assuming that there are no other problems, report as follows:

E2 Roof coverings

Slates cover the main sloping roof. Several slates are loose and cracked, allowing a leak to the room below. One slate is in danger of falling and is a safety hazard: CR3.

The pitched roof covering is in poor condition and causing damage to the building below (see F2). Although this can be temporarily repaired, it may be more cost-effective to replace the whole covering now.

To repair roofs safely and without damaging the roof covering, contractors will have to use appropriate access equipment (such as scaffolding, or hydraulic platforms). This can increase the cost of the work.

J3 Risks to people

E2: Roof coverings – loose.

APPLYING THE CONDITION RATING TO MULTIPLE BUILDING ELEMENTS

Well-structured reports usually begin with a summary to give readers an overview of what is to follow. In the case of the new HBR, a summary of the condition ratings is included in section C, immediately after the introduction.

Some surveyors worry that the reader will only read section C, and not bother looking at the rest of the report. This is always a danger, but I think it is a risk worth taking if the result is a more accessible report. If section C were to include condition ratings for all elements and sub-elements, it would extend over many pages and undermine the original intention. For example, a maximum of 27 elements could be condition rated; each one may be divided into three sub-elements (one for each condition rating) – giving a total of 81 condition ratings to fit in section C. Most readers would find this overwhelming.

To avoid this, the practice notes establish the principle of taking forward only *one* condition rating to section C. Here is a simple way to do this in a consistent fashion:

1. Identify elements and sub elements. Where an element consists of one or more different and distinct 'parts', it should be divided into appropriate 'sub-elements'. For example, a typical semi-detached inter-war dwelling could have: three

different types of roof (the main hipped and pitched roof, a roof over the front bay and a flat roof over the back addition); two chimneys (the larger chimney at the ridge serving the principal rooms and a single chimney from the kitchen area); two types of external wall (cavity walls to the main house, solid walls to the back addition); and two types of windows (the original single-glazed casement windows to the rear and uPVC double-glazed replacements to the front and side). All these are 'sub-elements' to be considered separately. This sub-division should be strictly limited to the generic parts only. For example, do not be tempted to further sub-divide the cavity wall into the front, rear, and sidewalls or the main roof into front, rear and hip slopes.

2. Apply a condition rating to the elements and sub-elements.
3. Establish the overall element rating. Where there are a number of different sub-elements and condition ratings, choose the 'worst case' to represent the element as a whole. This will then be shown in the coloured circle in the elemental text box in section C.

Although this approach keeps the summary section as simple as possible, many commentators are worried that this may give an unnecessarily negative impression of the property. For example, go back to the building described above, and imagine that the main roof is allocated CR1, the back addition roof is a CR2 and the small but leaking bay roof is given a CR3. Using the 'worst-case' principle, section C will show a CR3 under 'E2 Roof coverings', which could worry the reader – even though the cost of repairing the smaller bay roof may be relatively inexpensive.

Although this 'skewing' of the property's condition is a concern, it is partially ameliorated by the presence of the text box 'Surveyor's overall opinion of the property' in the same section. I think this also represents the main advantage the new HBR has over the old Home Condition Report: the HBR does not just rely on the condition ratings as the sole descriptor of the state of the building. Both in the Summary section (C) and in the text boxes of the individual elements themselves, you will be able to explain, justify, and provide additional advice that will help the reader place the condition in context.

CONDITION RATING OF MISCELLANEOUS ELEMENTS

Decorations

Decorations – either external or internal – do not have a separate element. Instead, they are assessed along with the element with which they are associated. In my view, this is a sensible approach, as removes the need to separate the decorations from the element they were applied to protect.

For example, if the decoration of timber window frames is satisfactory and there are no other defects or deficiencies, then 'normal maintenance' would be appropriate and a CR1 allocated. If the paintwork is beginning to peel and flake, a few of the window putties were cracked and missing, and if there are a few small areas of wet rot, then a CR2 could be suitable. But if there were large areas of wet rot, several casements did not open and a cracked pane of glass was not properly secured, the rating should be CR3. In the latter, it does not matter that the decorations were renewed only six months ago, a CR3 is still required.

Conservatory and porches – the new building element E7

This new category of building element has been created because these features are often expensive and sizeable additions to a dwelling. Rather than treat them as outbuildings, a separate element is justified. If they are not designed, installed and maintained properly, they can result in expensive remedial work, present safety hazards and cause legal problems. Here are some examples of assessments for these elements:

uPVC conservatory

Figure 5.15

This typical uPVC conservatory is constructed across the full width of this semi-detached house. Points to note:
- In order to be exempt from Building Regulations, the conservatory has to conform to a specific definition. If it does not, then it will need Building Regulations approval as if it were an extension.
- In conservation areas and on listed buildings, it may need planning approval.
- Is it being used appropriately? Some owners regard conservatories as an extension of the home. Many conservatories are turned into utility rooms and

even kitchens with cookers, fridges and washing machines.
- If the conservatory extends across the full width of the house it may restrict escape from the upper windows in the case of fire. Although this is not a mandatory requirement, you may want to make the client aware of this fact.
- Whether defined as a conservatory or not, any low-level glazing must conform to the Building Regulations standard (Part E), i.e. safety glass is required. Without that, it would be a hazard.
- Many owners take out the door and windows between the conservatory and the main home. This will increase heat loss and fuel bills, but it might also breach Building Regulations.
- Is the junction between the roof and the walls properly water-proofed? It is rare to see a cavity tray installed at the junction of the conservatory roof and house wall; without it, water could bypass any flashings and affect the conservatory room below.
- Has the structure been built over any foul or surface water drains? If yes, has this been properly approved and executed?

When assessing and allocating condition ratings to conservatories and porches, you should assess the feature as a whole, rather than split it into separate elements. For example, you should not split the conservatory into 'roof', 'walls' and 'floors' and include these under separate element headings – that would be too confusing. But if the conservatory or porch contains services, you should report these under the relevant elements in section G.

Assuming that it has got no serious problems and the lower glazing is marked as safety glass, a typical condition rating and report for the uPCV conservatory could be:

E7: Conservatories and porches

The conservatory to the rear of the house is made with uPVC frames and a combination of glass-double-glazed units to the walls and plastic sheets to the roof. The waterproofing at the junction of the main house and the conservatory roof (called the flashing) is loose: CR2.

The flashing needs to be repaired soon, because rainwater may damage the space below. Highly glazed conservatories will always lose a lot of heat, may suffer from condensation, and may affect the enjoyment of this space, especially during the winter when it may be too cold to use.

You should ask your legal adviser if the conservatory has a guarantee or a warranty from the installer (see section I). If these do not exist, you should ask an appropriately qualified person to assess the quality of the conservatory.

The timber 'sun room'

Figure 5.16

This older structure has been constructed across the rear of a semi-detached house. In addition to many of the points raised above, I would add the following:

- The structure should not be regarded as a 'conservatory' because not enough of the roof is glazed.
- It certainly does not meet the standards of an extension.
- The low-level glazing is almost certainly a hazard, as it is unlikely to be safety glass.

Consequently, this whole structure is likely to receive an adverse rating.

Built-in elements

In the HBR, built-in fittings (F6) are emphasised by being allocated to a distinct element. So, as well as the classic example of the fitted kitchen, this category should also include the utility room, fitted wardrobes in the bedroom and even specially constructed fittings in the home office.

In my view, anything that has been specially made or adapted for a particular purpose and permanently fixed into position should be assessed and reported under this section. The most important criteria will include functionality (whether it works), adequacy (whether it is big enough) and condition. This not only mean the doors and drawers but also watch out for dampness, wood rot and wood boring insects, as many kitchen fittings will be close to leaking plumbing and damp floors and walls.

Bathroom fittings – the new element F8

The new element, F8 Bathroom fittings, means that these should now be considered separately (i.e. not part of F6 Built-in fittings). This has been introduced because many modern homes have an en-suite, a separate bathroom, and a toilet downstairs, making these high value elements. For example, an associate of mine recently moved into a larger new house; when asked how many toilets it had, she replied 'eight', adding that, although she had lived there for nearly five months, three had yet to be used.

As with F6 Built-in fittings, the condition of the actual sanitary appliances should drive the assessment process and this is about the physical condition, not about modernity or fashion. If there are saleability issues, the report deals with these. Therefore, although an avocado bath suite in satisfactory condition must be given a CR1, you will be able to express your full opinions (providing they are professionally presented!) in the comments box.

The practice notes usefully identify those parts of the bathroom fittings to assess and report. The challenge is to identify where one element ends and F8 takes over. The overlaps include: G3 Water, G5 Water heating, and G6 Drainage. The simple rule to follow is that G3 Water and G5 Water heating finish at the point where the water pipes connect to the tap; and G6 Drainage ends where the waste pipe and trap connects to the waste fitting associated with the feature.

(a)

(b)

(c)

Figure 5.17

The bathroom fittings in this 28-year-old property (a) may not match modern preferences, but the surveyor should focus on their condition. Figure 5.17b shows a close-up of the bidet; and (c) is another close-up, showing the plastic waste pipe and trap (part of G6 Drainage), the hot water pipework (G5 Water heating) and the cold water pipework (G3 Water).

Insulation, ventilation and condensation

Under the previous version of the Homebuyer Report, I always felt these issues were included in C4 as a bit of an afterthought. In theory, this approach was sensible because the relative balance of insulation, ventilation and heating will determine whether the building suffers from condensation-related dampness. However, the previous practice notes included pipe insulation, under floor ventilation and a variety of other matters but no mention of heating. Under the new elemental HBR, these issues should be dealt with as follows:

Insulation:

- roof/loft space insulation – F1 Roof structure/space;
- wall insulation – E4 Main walls;
- floor insulation – F4 Floors;
- double glazing – E5 Windows;
- pipe and tank insulation – G3 Water (to prevent freezing), G4 Heating (to stop the condensate pipe from freezing) and G5 water heating (to reduce heat loss); and
- sound insulation – the element that is affected for example party walls and floors.

Ventilation:

- roof space ventilation – F1 Roof structure/space;
- under-floor ventilation – F4 Floors;
- general room ventilation – E5 Windows (for trickle vents and opening windows), but note that this can be slightly more complicated, as illustrated below.

Condensation-related defects

(a)

(b)

(c)
Figure 5.18

Figure 5.18a–c shows the bathroom ceiling of a house built in the 1980s. Figure 5.18a shows mould growing on the ceiling, suggesting a condensation problem (assume that the roof is not leaking and there are no pipes in the vicinity). The bathroom is heated by a radiator from a gas-fired central heating system, the window (Fig. 5.18b) is double-glazed but is not fitted with a trickle ventilator, the room does not have any other form of ventilation. An inspection of the loft space (c) above reveals that the thermal insulation has been moved during a previous repair and not replaced.

Although the problem with the loft insulation has created an intense 'cold spot' where the moist air in the bathroom has condensed on the underside, I do not think it is a simple matter of putting back the insulation. The extent of the problem suggests an imbalance in the environmental conditions in the room.

Because condensation is a dynamic defect that is related to the way a space is used as well as how the building is constructed, it does not sit ideally within an elemental report. Although the bathroom appears adequately heated, the lack of ventilation will result in regular high moisture levels enhancing the cold spot condensation. So how can this complex problem be condition rated and reported? Here are my thoughts:

- The problem is primarily affecting the ceiling, so the main rating should be applied to F2.
- Although the mould is clearly visible, I am not convinced that it is a serious or urgent problem. To make sure, I will apply the questions from the condition rating protocol (described earlier in this chapter):

Question A – Although mould is growing on the ceiling, I do not think it is impairing its function. Consequently, I ticked the 'no' response.

Question B – The mould could begin to grow on the walls but at the moment it is not affecting any other elements so this response was a 'no' as well.

Question C – The stability of the ceiling is not threatened, so this is a 'no'.

Question D – The health and safety question is a challenging one. Mould spores and damp environments are injurious to health, especially for people with allergies. In this case, is there enough mould and would people be exposed to it for long enough periods? I am no expert but I will have to take a judgement. Consequently, I would say not. If this were a bedroom, it would be a different matter. I have also ignored the scare stories about 'toxic mould' as I do not think the evidence shows it is as big a threat as the specialist mould removal contractors claim.

Question E – Following on from the logic in question D, I do not think this is an urgent repair.

Question F – Further investigations are not required.

As a result of this process, I would rate this problem CR2. Nevertheless, as this is a complex issue, I would include the following advice in the comments box of F2 Ceilings:

- The ceiling surface should be treated and sealed.
- The thermal insulation should be refitted in the loft (cross reference to F1).
- To help prevent the problem in the future, additional background ventilation should be installed in the bathroom.

It is beyond the scope of the HBR to provide specific details of the repair, but I think it is right to point the potential new owner in the right direction.

Garages, outbuildings and grounds

Section H in the new HBR is quite different to previous report structures. It places more emphasis on the external features, for a number of reasons:

- There is some evidence that surveyors often lose concentration when they get outside. After 90 minutes of focusing on the property, many seem happy with a brief walk around the gardens.
- Some insurers have noticed the number of claims associated with garages have increased. Higher levels of car ownership and a lack of on-street parking in some towns and cities now means that garages are both very useful and valuable.
- Some external features are expensive to repair, but if they are not properly maintained, the value of a property could be affected. For example, a large retaining wall in a hilly neighbourhood can cost tens of thousands to rebuild and create considerable legal tangles.

To emphasise these issues, the section has been split into three sections: H1 Garages, H2 Other (to include other permanent outbuildings), and H3 General (aspects associated with the grounds). Of these, only H1 and H2 are condition rated. Nevertheless, where an external feature is likely to affect the value of the property, it can be 'promoted' into section H2 and condition rated.

This may sound a little vague but I think it provides useful flexibility. Many external features are secondary to potential buyers when they are making their decision to purchase and so the condition rating of minor features should be avoided. To use the example I described above: a 1.75m high retaining wall on the verge of collapse will cost a small fortune to rebuild, but without that work, it could also be very dangerous. By moving this feature into H2, you can use the condition rating system for what it is designed; to emphasise a serious issue. Otherwise if this was merely described in H3, the client might not see it as an important issue.

As you know, garages come in all shapes and sizes. These buildings can be very poorly built, never repaired and often used for some unusual purposes. The practice notes identify typical features that you should look out for:

- general condition;
- functionality – is it large enough for a modern car;
- unauthorised use – many are converted into workshops or places of work;
- services – a classic place for DIY electrics and creative but dangerous heating systems; and
- planning/listed building/conservation area issues (the rules in some areas can be very strict).

Here are a few examples and how I allocated the condition rating.

The 1940s garage

(a)

(b)

(c)
Figure 5.19

The garage shown in Figure 5.19a–c is along the side of an inter-war semi-detached property, and probably dates back to the 1940s or 50s. It would have been ideal for an Austin A40, but is a bit slim for a modern Audi saloon.

Most commentators suggest the absolute minimum internal dimensions for a garage are 3m × 5m but 3m × 6m would fit a greater range of cars. Any less than this and it simply would not meet the functional requirements. In this case, although I could not get inside (it was padlocked and the agent did not have the key), there are a number of features that concerned me:

- It is constructed of materials that probably contain asbestos. Although I cannot be sure without having the material tested, the roof is likely to be made of asbestos cement and the walls of 'asbestolux' or some other form of wallboard.
- Closer inspection (but no too close) reveals that the walls and roof are in poor condition. This means that the building should not be used because of possible asbestos fibres. It will have to be made safe or demolished. This is a specialist operation and will be costly.
- As Fig. 5.19c shows, there is a unique method of getting a cable from the main house to the garage. Although I am not an electrician, I do not think a plastic rubbish bag is a suitable method of insulating a live cable. This will be well below current standards.

The garage is clearly in need of repair or replacement. I should go through the questions in Stage two of my protocol (as set out at the start of this chapter) but to save space here, I have just listed those I would answer 'yes' to:

Question A – The size of the garage impairs the function of the element.

Question D – The electrical system and asbestos content will threaten the health and safety of the building users.

Question E – Urgent repair: I had to think carefully about this aspect because although the garage is almost certainly contaminated by asbestos fibres, as long as the users do NOT go in there, it is not an urgent problem. However, that electrical cable is worrying and needs to be resolved straight away.

In respect of the other questions, although the garage is a poor condition, I do not think it is structurally unsound (Question C); it will not cause problems in other building elements (Question B) and no further investigated is needed (Question F) so I answered 'no' for all of these.

Looking at these responses, CR3 would be appropriate, as initially thought. However, because I could not get inside to carry out an inspection, I should clearly state this in the 'I could not inspect …' box at the start of this section of the report. Before I move on to the next example, I would like to revisit the 'further investigation' issue. Resolving the problems of this garage will involve contractors with specialist knowledge who may have to investigate the true nature of the material. But for the purposes of condition rating and producing the HBR, the surveyor should have all the knowledge to make the initial decision.

The 1980s attached garage

Figure 5.20

The garage shown in Figure 5.20 is attached to a detached house built in 1981 and although the garage is a separate entity, it was built at the same time as the dwelling using the same construction methods.

The question that needs to be resolved is whether the garage should be assessed within the elemental structure of the main property or condition rated separately under H1. This will always be a matter of judgement, but I think that this should be assessed under H1 because of its distinctive and separate nature. However, there are a number of 'what to put where' issues:

- The separate elements of the garage roof, walls and floors should all be assessed together and included in H1.
- The consumer unit for the whole property is in the garage. This should be included in G1 Electricity in the normal way.
- There is a sink in the garage and, in theory, this can be assessed across four different elements: F6 Built-in fittings (for the sink); G3 Water; G5 Water heating; G6 Drainage. Splitting this sink across all these elements simply does make sense and it would be very difficult for the client to follow. Consequently, you should report on aspects of it under H1.

The following principles may help you decide how to report on the garage

- If the garage is completely separate to the main dwelling, report on all aspects under H1.
- If it is simply attached to the dwelling (shares a common wall), it should be reported under H1, but where the garage contains elements that mainly serve the main property, these should be reported under the appropriate building element within the main report. Examples include the service head, electricity meters and consumer unit to the whole dwelling, a sealed inspection chamber, etc.
- If the garage is mainly integral to the dwelling, it should be rated and reported as if it is part of the house (see below).

These are not strict rules, as they may not fit every circumstance, but you should structure your report logically so that it is easy for you to produce and for the client to read.

Integral garage

Essentially, the entire garage shown in Figure 5.21a–c is contained within the structural framework of the dwelling. Therefore it should be assessed as part of it. However, the garage is approximately 2.3m wide and this is further reduced because of the soil and vent pipe. To make sure the reader of the report is aware of this potential restriction, you could use H1 to assess the garage's functionality, and in this case, I would mark this CR3.

(a)

(b)

(c)
Figure 5.21

Outbuildings

As described in the practice notes, section H includes other permanent outbuilding(s) that are not attached to the main dwelling and must be condition rated and reported. Examples include:
- summer houses;
- substantial greenhouses;
- follies;
- leisure buildings but not the leisure facilities inside (for example, swimming pools, saunas, fitness gyms, etc.).

As with garages, you should not split the building into separate building elements, but instead assess and rate it as a whole. In some cases, this could include substantial and unusual structures. The following points may also help:

- Where the building is attached and clearly part of the main dwelling, consider incorporating it within the elemental structure as discussed under 'Garages'.
- A 'substantial greenhouse' is taken to mean a structure that involved some level of building work to create. Examples could include foundations and concrete floor beds, dwarf brick walls on which the glazed walls are fixed, etc. If the greenhouse cannot be simply dismantled, loaded into a removal lorry, and taken away, then you should apply a condition rating. Where it is possible to take the greenhouse down (but the current owner still plans to leave it), in my view you should still describe it, but *not* apply a condition rating. This keeps within the spirit of the HBR.

Figure 5.22 shows a brick-built toilet in the rear garden of an older terrace. Although this is attached to the dwelling, it is clearly not part of it, and should be reported under H2. In Figure 5.23, the free-standing garden sun-room should be assessed separately under H2, if it is to remain. Believe it or not, the building shown in Figure 5.24(a) and (b) was discovered at the end of the garden belonging to a bottom-of-the-market semi-detached non-traditional concrete house. Set in a big plot, this is not your normal 'garden shed'. An internal view of the building shown in (b) suggests that there are some legal and planning issues to be resolved here.

Figure 5.22

Figure 5.23

Figure 5.24(a)

Figure 5.24(b)

Other external features (section H3)

In section H3, the practice notes identify a number of external elements that should only be reported. Although you should identify, describe, and assess these, you should NOT apply a condition rating. This avoids the silliness of having to give condition ratings to small patios, minor access paths and other garden features.

However, where there are more substantial defective external features, which could affect the value of the property, then they should be taken out of this section, condition rated and reported in H2. Examples include:

- retaining walls in danger of collapsing;
- deeply sunken paths or driveway that could be expensive to repair;
- dilapidated boundary wall or fence that has cost or safety implications.

There must be a clear impact on value so inclusion in this section is likely to be the exception rather than the rule.

Figure 5.25a shows a high retaining wall that holds up the garden of the semi-detached house at the higher level. The wall is in four different sections and I doubt

Figure 5.26

that any have been properly designed. There would be serious implications if a wall like this were to collapse – which it did (b), the next day. This is a very dramatic example. The cost of rebuilding the wall would be tens of thousands of pounds, and so clearly would be a matter to include in section H2.

The timber boundary fence shown in Figure 5.26 would be a matter to include in H3. However, the fence is 53m long; so perhaps that could be a reason to transfer the feature to H2 and condition rate it? I think so.

Figure 5.25(a)

Figure 5.25(b)

Trees

Section H3 also includes trees. Where these are not the cause of problems, then they should be described and, if appropriate, give general maintenance advice helpful to your client. On the other hand, if the trees are presenting a risk or causing a problem, report this in H2 and apply an appropriate condition rating. For example:

- in an area of clay sub-soils, a vigorously growing tree with a large mature size located within influencing distance of the foundations of the property;
- a tree growing directly over the line of an existing shallow drain where the roots may cause damage; and
- where the branches of a large tree extend over parts of the dwelling and are a danger for passers by and to the property itself.

Within the report, you may also want to:

- cross reference to the element that it is affecting;
- include the matter in section J Risks.

Other issues could include:

- listed building/conservation areas (cross-reference to section I Legal issues);
- safety issues (cross-reference to section J Risks).

CONDITION RATING OF 'RISKS TO THE BUILDING' (SECTION J1)

As all surveyors know, structural movement, dampness, timber problems and condensation, are the defects that can cause some of the biggest problems in a property. In the previous version of the report, these were emphasised in their own section at the beginning of the report. However, the new HBR is an 'elemental report', so this distinction is no longer possible because structural movement, dampness and timber problems are defects not 'elements'. Consequently you will have to capture these problems by allocating condition ratings and reporting on them within the elements that are affected.

This will require a slight change in approach, because these problems may affect a number of elements. To illustrate this, consider the classic case of structural movement to the corner of an inter-war semi-detached property. Let's assume the sub-soil is shrinkable clay and the problem has been caused by a willow tree growing in the neighbouring property. The combined foul and surface water drain shared with the neighbour runs between the tree and the dwelling itself. Could anything more go wrong? It may sound like I have created an unlikely scenario but it is not uncommon. Focusing on how this should be covered in the report, it may affect a number of different elements. I have made a number of assumptions for the sake of illustration:

E4 Main walls – The shrinkage of the sub-soil has undermined the foundations causing the wall above to crack. Because this is a serious defect that will also require further investigation, a CR3 is likely to be appropriate.

E5 Windows and E6 Outside doors – The movement in the walls has caused the openings to distort and the doors and windows to bind against their frames. Although not urgent or serious, these will need easing once the structure has been stabilised and a CR2 would be suitable.

F3 Internal walls and partitions – The same movement has also distorted internal walls and door openings, cracking the plaster and preventing the door from shutting properly. Assuming these repairs can wait until the problem has been resolved, this would be a CR2.

F4 Floors – If the movement has caused the solid ground floors to subside, although no repair may be required (not economically justified), you may want to report the problem to the client for reassurance and allocate a CR1.

G6 Drainage – Movement in the sub-soil may have caused the drain to crack allowing the roots of the tree to enter the pipes and cause a blockage. This is a serious defect that could be urgent if blocked and may require further investigation. Therefore a CR3 would be appropriate.

H2 Trees – Because of the seriousness of the problem, you may want to move the matter into H2 so you can apply a CR3 rating to the tree.

Section I: Issues for your legal adviser – Because the tree is in the neighbour's garden, they may be responsible for this repair. There may also be legal issues around shared drains, insurance cover, etc.

Quite a list! Because only part of the problem may be explained under each element, section J1 has been used to give the reader of the report an overview of the problem. Here you should only include a brief description of the problem and list the elements that it affects. Following my assumptions above, an appropriate clause could include:

J1: Risks to the building
Structural Movement – The house has been affected by structural movement. Please see E4, E5, F3, F4, G6, H2 and I3.

The same approach should be taken with the other cross-elemental defects including all forms of dampness and timber defects. This approach could also apply to properties in Cornwall if they are built of Mundic concrete or if a defective non-traditional system has been used.

FACTORS THAT CAN AFFECT THE VALUE OF THE PROPERTY BUT ARE NOT CONDITION RATED

Although the elemental structure of the report will capture most of the problems that affect the property, there will be a number of influences that fall outside this system. For these problems you will not be able to use the condition rating system to focus your client's attentions to the problem, so you must make sure you use other parts of the report instead. Here are some of these issues and suggestions of how they can be covered in the report:

- **radon** – should be included under section D Local environment and under section J3 Risks to people;
- **electro-magnetic fields** – as for radon;
- **mining activity** – include under section D Local environment and under section J2 Risks to grounds;
- **contaminated land/landfill** – as for mining activity;
- **flooding** – as for mining activity;
- **major nuisances** – (noisy or offensive industrial production, local nightclubs, etc.) should be included under section D Location and under section J4 Other;

- **flight paths, major roads, railway line** – as for major nuisances.

You may also want to reflect the issue in the space allotted in section C to 'Surveyor's overall opinion of the property', and of course take account of the matter when you are producing your valuation. By using the report in this way, you will be able to highlight those issues that fall outside of the rating framework.

MARKET VALUE AND CONDITION RATING

Valuation is not one of my core competencies and the focus of this book is on the process of inspecting, assessing and reporting on the property. Nevertheless, the HBR has a mandatory valuation component. Rather than trying to tell my grandparents how to suck eggs, I will not explain how to value residential properties. Instead, I will focus on the new feature of the HBR; the impact of the condition ratings on market value.

The practice notes presents a suitable methodology: 'The surveyor must assess the current Market Value after applying the condition ratings' [paragraph 4.6, page 18]. This is a sensible approach because an objective view of the condition of that particular property will help 'fine tune' its market value. The practice notes goes on to articulate the relationship between condition ratings and market value:

- Elements with a condition rating 1 will not have an impact on the Market Value of the property;
- An element with a condition rating 2 will not generally impact on the Market Value unless there is an accumulation of items with this rating that would increase costs to a limit where the surveyor judges that there would be an impact on Market Value;
- Elements with a condition rating 3 would generally have an impact on Market Value.

This is a good starting point and it would be wonderful if there was such a direct relationship between condition and value. However, the real world is slightly more complicated. Consider the following:

- In a falling market where the cost of building work remains stable, condition will have an increasing impact on value.
- In my neighbourhood in Sheffield, a CR3 defect that costs £1000 to put right would be a deal breaker. In Hampstead it would not be noticed.
- There are issues that will affect market value but will not be captured by the condition rating process. Location and environment issues are typical examples.

To illustrate this more clearly, consider the following examples:

Dangerous electrics – CR3 but no effect on value

Figure 5.27

This may not be the most dangerous electrical fitting you have come across but it deserves a CR3 as a safety hazard. On its own it is unlikely to affect value, but if it is part of a substandard electrical installation it might – especially if the property is at the bottom end of its market. In an area of better quality houses the valuation would be unaltered.

Major structural movement – CR3 and impacts on value

Figure 5.28

You should never assess structural movement based only on the cracks it produces, but sometimes the visual evidence is simply enough. Whatever the context, this is a serious defect and will affect the value of most properties.

Overhead pylons – no condition ratings but impact on value

Figure 5.29

This photograph shows high-voltage power lines passing directly over a semi-detached house that is in otherwise satisfactory condition. This property may have no CR3s because the power lines are not an issue captured by condition ratings. Instead it will be reported under: section C – overall opinion; section D – the location, and local environment; and section J3 – Risks to people.

Although there is still conflicting evidence about the health effects of electro-magnetic fields, most buyers would be concerned about this and surveyors are taking account of this in their valuations.

Arguably this is another advantage with the condition rating system; it could assist in the valuation process. Although you must look at a whole range of factors, the range of different condition ratings in the summary will give a quick snapshot of the property's condition.

CONDITION RATING BUILDING ELEMENTS IN FLATS

As discussed in chapter 3, the inspection routine for flats has not changed. The approach to condition rating is also the same, but the shared nature of some building elements makes this process a little different. To help with your decision making, this section describes an approach that you may find useful.

Elements that are exclusive to the inspected flat

Many building elements will be wholly contained within the flat, for example:

- the floor (in the lowest flat);
- the ceiling and roof (of the uppermost flat);
- walls and partitions (F3), ceilings (F2), built-in fittings (F6), woodwork (F7), bathroom fittings (F8);
- all services that are not shared.

All parts of these elements can normally be seen and assessed, so the usual condition rating methodology can be applied and the matter reported under the usual element headings.

In these cases, the report is constructed in exactly the same way as described earlier in this chapter.

Elements that are shared with the block

Some of the shared elements will need to be condition rated, and this is a 'grey area' where you will have to make a judgement. This category covers instances where most of the element can be evaluated but you have to use your knowledge and experience to assess the condition of the rest. For example:

- In a small residential block, the external walls (E4) are shared. In this case, you might be able to see the flat's walls both inside and out and the rest of the walls from the outside only. If signs or symptoms of defects are clear, you could allocate a condition rating to the wall element as a whole, while clearly stating the rating is based on a restricted inspection.
- The windows (E5) to the flat can usually be fully assessed, but those to the other dwellings can be seen only from the outside. The approach described for the external walls can be applied here.
- The rainwater pipes and gutters (E3), whether shared or not, can usually be assessed in the same manner as an individual property.

The restrictions to your inspection should always be clearly stated in the appropriate box at the start of sections E, F, G and H, but the rest of the report should be completed as normal.

> **E Outside the property**
>
> [In the box for explaining restrictions on the inspection ...]
> My inspection included the external features of the block containing the property and not the other three blocks on the site. I inspected the internal faces of the outside wall in this flat only.

> **E4 Walls**
>
> The external walls are made of brick with an air gap between the inside and outside faces (called a cavity wall). The mortar between the bricks (the pointing) has eroded. CR2
>
> The pointing should be replaced soon. You should ask your legal adviser about the responsibility for the cost of repairs to the external and shared areas of the estate (see section I).

Shared elements where an assessment cannot be made

This typically includes elements where so much is concealed that a speculative conditional assessment is not possible. Alternatively, the element exhibits a 'trail of suspicion' that cannot be followed far enough to come to a view. For example:

- The external walls (E4) are part of a larger block and consist of a number of different walling types, some of which may not be observable from the ground. In this case, the walling to the subject flat has only a small influence on the condition of the walls to the block as a whole.
- The windows (E5) can be accurately assessed for the flat you are inspecting, but what if the flat is part of a block of 100 or more, each with six or seven windows? What if the flat has its windows on the sheltered side of the block but those on the elevation facing the weather are beginning to rot? There is a danger of allocating a lower condition rating based on what you find in the flat and missing the poor state of the remainder.
- The rainwater pipes and gutters (E3) on larger blocks can be partially or wholly concealed preventing an accurate assessment.

In these cases, the decision will follow that described for 'further investigations' on page 40. Here the fourth condition rating of 'Not inspected' (NI) is appropriate, because there is simply not enough information on which to base a decision. In all NI cases, you will still need to make a full report so the client can still see what the potential problems could be.

Where there are clear signs that a serious defect may exist (for example, the walls are cracked or the building is affected by extensive dampness from an unknown source), allocating a CR3 is appropriate because you know that something is wrong – it affects the subject flat and it requires further investigation.

Services

Services that are normally reported in the HBR including Electricity (G1), Gas/Oil (G2), Water (G3), Heating (G4), Water heating (G5) and Drainage (G6). Where these are wholly confined to the subject flat only, then it can be assessed in the usual way. However, where part or all of the service system is shared with two or more flats, it will be impossible to allocate a rating. Typical examples could include:

- heating from a communal system;
- water supply from a large central storage vessel;
- internal soil and vent pipes that receive the foul water drainage from a number of flats.

In these situations, you should allocate the condition rating NI, because the assessment is complex and very technical.

Services that are specific to flats

The HBR includes a specific element in section G7 called 'Common services', such as:

- refuse chutes, bin stores, etc;
- lifts – both passenger and goods;
- entry-phone systems;
- CCTV;
- lighting and heating in common areas.

The practice notes advise that these services require 'general comment only'. Therefore, no attempt should be made to condition rate these, and this section of the report should be treated in the same way as section H3 Grounds (see below).

Using the estate shown in Figures 3.12, 3.13 and 5.30(a) and (b) as an example, and assuming it has rubbish chutes and CCTV cameras at various points around the blocks, a report could include:

G7 Common services

The estate has a number of services common to all flats:

- Refuse chutes – There is a refuse chute on each floor adjacent to the central staircase. This discharges into large steel rubbish bins in store cupboards on the ground floor. These are generally well maintained, although the doors to the bin store were not locked making it vulnerable to vandalism.
- CCTV system – I noticed a number of close circuit television cameras around the estate. I am unable to comment on whether these work effectively.

You should ask your legal adviser to investigate who is responsible for these facilities (see section I).

External shared areas

This typically includes:

- external staircases;
- outbuildings;
- garages;
- parking provision;
- roads and paths;
- boundaries;
- retaining walls, etc.

Unlike self-contained dwellings, the common areas of flats are not easy to condition rate because of complex legal ownership issues. Consequently, I think you should sparingly 'promote' an H3 feature to H2 so it can be condition rated, ideally reserving this action for safety hazards. Otherwise leave these aspects in H3, as the following example demonstrates.

Figure 5.30(a)

Figure 5.30(b)

Figure 5.30(a) shows the garages and associated roadway on an estate. Because the flat may not have exclusive use of a garage, these elements should not be condition rated, only described. Figure 5.30(b) shows the parking area of the same estate, in particular the condition of the surfacing. This is the most obvious feature to report, but do not forget the trees. And what about the very high retaining structure on the left? If that collapses and the estate is responsible, flat owners could face a very large bill.

H3 General

The estate includes a number of features:

- Communal garages – There are approximately 20 garages towards the north side of the estate. These are made of precast concrete panels with cement-based corrugated roof sheets. These may contain asbestos. You should ask your legal adviser if the estate's management company have an appropriate asbestos management plan in place (see section I).
- Estate roads and car parking areas – There are a number of parking areas and linking estate roads that are surfaced with tarmacadam. These are generally in a poor condition with many holes and evidence of previous patch repairs. These areas will need to be resurfaced in the near future.
- Trees and other landscaped areas – The landscaped areas of the estate are covered with lawns and the occasional mature tree. Some of the trees are overgrown and need to be properly maintained to avoid future safety hazards.
- Retaining wall – The boundary to the south side of the car park is defined by a steeply sloping rock face approximately 15m high. This is overgrown with vigorously growing saplings that may threaten the stability of parts of the cliff face. You should ask your legal adviser if you will have any responsibility for the maintenance of this feature. If yes, you should ask an appropriately qualified person to report on its stability and condition.

6. Writing the report

Research carried out by RICS and my own anecdotal feedback suggests that many surveyors' reports are not as effective as they could be. Here are some commonly cited problems:

- full of jargon and long technical descriptions of parts of the building or general explanations of different defect types;
- caveats or defensive statements that either restate the limitations in the conditions of engagement or introduce new restrictions that go beyond those stated;
- a general 'passive' style of reporting that gives an overall negative impression to the reader.

In my various roles in the property sector, for one reason or another, I review a large number of reports. Although I cannot describe myself as a master of the art (I still struggle with basic grammar – a product of the 1960s' education system), it is easy to stand back and judge that many do not read well. For example, I found the following phrases in one report:

> '… appears to be in a satisfactory condition at present …';
>
> '… nothing has been found to suggest that there may be a significant defect …';
>
> '… no significant defects were noted …'.

These are common and although I understand why they are used, they can give the reader an overall negative impression. The Plain English Campaign (www.plainenglish.co.uk) recommends that our writing should be 'active', sounding crisp and professional, rather than 'passive' that is stuffy and bureaucratic.

Even though it is beyond the scope of this book (and possibly the competence of the author) to cover report-writing techniques in detail, there are a few methods that may help you review your own approach. They are not 'magic bullets' that will cure all problems but I have found they improved my technique.

Consider the following example, which is taken from a report that describes the construction of the rear part of the property being inspected.

> *There are a series of regular header bricks provided to the offshoot walls and as such, taking into account the width of the walls these are referred to as 'snapped headers' whereby these provide the bond between the inner and outer leaf brickwork to the cavity walls whereas more modern construction methods are to provide cavity metal wall ties.*

What a monster of a sentence! In my view, the main problem is that the sentence is purely descriptive; it offers no advice. Does the client really want to know this? In addition, I am not sure it makes any technical sense.

Putting these technical concerns aside, one method of checking your text is to use the free 'Drivel Defence' software on the Plain English Campaign's website. Normally located under the 'free guide' section of the site, it enables users to check their use of plain English. It is simple to use and if you 'cut' and 'paste' your plain text into the software, it will carry out a basic analysis. I analysed the above example and the software produced the following information:

Total sentences with 20 or more words = 1

Longest sentence has 59 words

Total potential alternative words = 6

Total words = 59

Total sentences = 1

It also suggested alternative words that could be used, which are automatically selected from the Plain English Campaign's own guide, *The A – Z of alternative words*. The software highlights those words that could be changed and recommends alternatives and these are highlighted in bold below:

> *There are a series of regular header bricks **provide**d to the offshoot walls and as **such**, taking into account the width of the walls these are **referred to as** 'snapped headers' whereby these **provide** the bond between the inner and outer leaf brickwork to the cavity walls **whereas** more modern construction methods are to **provide** cavity metal wall ties.*

The Plain English Campaign's recommended alternatives include:

provide → give

referred to as → called

such → the, this, that

whereas → but

This is a good start, because it identifies a long sentence and suggests more suitable words. Using some of these simple adjustments, I have re-written the sentence:

> *The cavity walls to the rear of the house (called an offshoot) are made of brick. The inner and outer parts of the cavity wall are held together with bricks instead of metal wall ties, which are used in modern cavity walls.*

Total sentences with 20 or more words = 1

Longest sentence has 23 words

Total potential alternative words = 0

Total words = 42

Total sentences = 2

This is getting much better, but does the client need all of this description? Are they really interested in the differences between the methods of tying the walls together, or do they just want to know if it satisfactory?

I suspect the latter. To be fair to the surveyor who wrote the original version, describing how part of a building is constructed can be a useful first step in explaining a complex defect, but many practitioners manage to produce prose that would be fine in a construction technology book, but is not appropriate for non-specialist readers. The HomeBuyer Report, like the HSV before it, should be an economic report and so we need to communicate clearly and concisely.

There are other computer-based tools that can assist writers, and the most accessible of these is the spelling and grammar checking facility that is built into Microsoft's programs, notably Word. As described in the *Home Inspector's Handbook* (Parnham and Rispin 2005, page 77), these use a number of techniques that can help you evaluate and improve your writing.

Not all people are comfortable with using such a mechanistic approach, however. For example, a book editor once told me that spelling and grammar-checking software are notorious for mangling English,

---

Handling technical terms

The general commentary of the practice notes explains that the HBR is concise in fact as well as in theory '… *and to avoid confusing the client with distractions (such as relevant and unhelpful details and surveyor's jargon, which can be incomprehensible and off-putting to the layperson)*'.

Over the years, I have been fascinated by the way surveyors use terms that are familiar to us but a complete mystery to our clients. Here are a few of my favourites extracted directly from real reports:

- 'The benching in the inspection chamber is in very poor condition' – I bet the reader wondered why a seat had been installed in the drains in the first place.
- 'The property has suspended timber floors' – Does this mean they are hanging by cables?
- 'The front purlin has a number of significant shakes' – What is a purlin and why is it shaking so much?
- 'There is evidence of significant humping over the party wall' – I could not possibly comment on this description of an undulating roof covering!

These may be cheap jibes, but sometimes we forget how particular our technical language is. Consequently, you should avoid using such terms wherever possible.

To help with this, a 'cut-away' house diagram is part of the standard report package and this locates and names a range of typical building parts. Although this is helpful, it still does not explain what those parts actually do. For example, the reader will know where the flaunching and the soakaways are but not their functions.

On the other hand, knowing the technical names of parts of the building can be helpful for clients. For example, try to imagine asking a builder to repair your 'horizontal water-proof layer in the wall'; it will get a colourful response. The following approach might be useful:

- describe the building component in layperson terms; and
- use the technical term in parenthesis after the lay description.

In danger of being hoisted by my own petard (or should I say lifted by my own explosive device), I have rewritten the report extracts I described above:

- The surface across the base of the inspection chamber (called benching) is in a poor condition.
- The property has timber floors.
- The large timber beams that support the front and rear roof slopes (called the purlins) are split in a number of places.
- The roof covering is uneven where it passes over the wall shared with the neighbouring property (called the party wall).

I am sure that you can find fault with these but you should develop your own clear style that lay readers can understand.

adding that: 'Spelling chequers are knot infallible'. The editor also commented that mathematical formulae for measuring and assessing writing styles (especially those that judge sentence length) may be relevant for tabloid newspaper editors, but are not so appropriate for professionals, including surveyors writing HomeBuyer Reports. She recommended that we should always use a recently published dictionary and, if we are writing a report for a householder, why not imagine that the reader is your friend, neighbour, or a member of your family. This will help you check that all the explanations are clear, concise and helpful, without over-simplifying or patronising.

Personally, the techniques I mentioned above have helped me reflect on my own writing style. They have enabled me to look at my reports in an objective way and to adjust my approach. I would not recommend that you use these software programs all the time, because you need to use your own natural style. However, using them to analyse one or two reports can help you reflect on your own style.

Whether you are new to the sector or your hair is as grey as mine, writing styles can always be improved – and the launch of this new HBR product is a good time to reflect on your own style.

HOW TO COMPLETE THE COMMENT BOXES

Each section of the report lists the various building elements that are to be inspected, and an expandable space is provided below each heading so that you can explain the condition, justify your condition rating decision and offer any appropriate advice your client may find helpful.

As I write this book, and with the new HBR just launched, software producers are planning to offer a variety of report-writing packages that will populate the comment boxes with standard phrases. However, many people prefer to use their own text and phrases. In this case, you should structure your comments as described below. This will help to ensure a consistent approach is adopted by surveyors across the country.

For each condition rated element or sub-element there should be two parts: the descriptive and advisory sections.

Descriptive section

The first part of the report should be a brief, objective description of what you have found during your inspection. Do not include any 'advice' in this action. The description should consists of the following parts:

- Describe the construction and location of the element or sub-element. This should be short and simple avoiding technical jargon wherever possible.
- Describe the type of defect in broad terms. For example:

  – the actual part that has something wrong with it (the slates, the flat roof covering, the brickwork, etc.);
  – what is wrong with it (it is broken, leaking, missing, rotten, etc.);
  – the magnitude of the problem (it is completely broken, largely missing, completely rotten, etc.).

At the end of this paragraph, simply list the condition rating and highlight in bold. You should use one of the following phases:

Condition rating 1

Condition rating 2

Condition rating 3

Condition rating 3 (further investigation)

Not inspected.

There is no need to repeat the full definition of the categories as they are explained in section B of the report.

Advisory section

Unlike the Home Condition Report, the HBR allows you to provide the client with additional information and in my view this provides the 'added value' of the product. However, as it is still an 'economic' service, the section should be as short as possible. The following guidance may be helpful:

- Add more information about what needs to be done and by when. Although this is partially described in the condition rating definitions, you should provide helpful information that will put the client in the right 'ball-park'. This could include:

  – what needs to be done (but it must stop well short of a detailed specification, though you could sketch out the extent of the repair or replacement work);
  – by when (match the time scales to the condition rating categories using words such as 'soon', 'as soon as possible', or 'now').

- Include other factors that will help the client understand the full extent of the work. For example, the:

  – impact of listed building status and the use of traditional building techniques;
  – need to obtain a party wall agreement;
  – implications of contractors working at height;
  – implications of possible asbestos content or other particular risks, etc.

- You must not give any detailed information on costs, but where these other factors will make it more expensive, consider adding a phrase like 'this can be costly'.
- Where the on-going maintenance or repair would be relatively expensive, you may want to advise that total replacement will be an economical choice.

I regard the HBR as a report on a property provided by an experienced professional who has been appointed by a specific client. This means, I think, that it is important to establish sector-wide standards to make sure home buyers know what to expect. However, within those parameters, there is scope for personal choice, specialism and even personality to be expressed. For example, a close colleague who has written more HSVs than I have sold books (which is no challenge) prefers a shorter, punchier style than my own. His clients (who use him again and again) prefer it that way, but others may find his reports a little abrupt. The principle here is to develop your own style that conforms to the practice notes and meets your clients needs.

OTHER REPORT-WRITING ISSUES

How long should the report be?

There is no prescription on length, apart from the practice notes reminding us that it is an 'economic' service. I have often asked surveyors at training seminars and the reported average is in the region of 3000–3500 words. The sample HBR in the appendix is approximately 4250 words long.

Reporting on further investigations

Where you have recommended further investigations, you should not specify the type of organisation or individual who should do this. This is because 'competent person' schemes and other regulatory initiatives have increased the numbers of 'specialists'. For example, there are a total of eight schemes listed under the 'electrical safety in dwellings' heading on the competent persons section of the Department for Communities and Local Government (CLG) website, including NAPIT, NICEIC, APHC, CORGI, and even OFTEC. Instead of even-handedly including all of these, you should keep it simple and include phrases like 'appropriately qualified and experienced professionals' or 'appropriately qualified and experienced professional contractors'. Not only does this keep your report short, it avoids confusing your client with unnecessary detail.

If you identify a problem that needs further investigation, your client may need help in deciding who to use, but that is another professional service. Although it may be difficult to achieve in these challenging economic times, you could provide this as 'Additional advice' and they should pay for it!

Ordering the text

The HBR is tightly structured, and the order and content of the elements cannot be altered. However, within the elemental text boxes there is some discretion. As I discussed on page 65, just one condition rating (usually the worst) for each element can be presented in the summary in section C.

Many commentators are concerned this may 'skew' the reader's opinion of the property; a view that can only be investigated after the HBR has been in widespread use for a period of time. Until this approach is evaluated, careful structuring of the text can help to produce a more balanced view.

To illustrate this approach, consider the property used for roof covering C on page 49. In total, this house has three different roof types:

- a slate-covered pitched roof;
- felted flat roofs to the front and rear dormers; and
- lead-covered flat roofs to the front and rear bay windows.

Because these are distinctive types of roofs (called sub-elements), each one will be allocated a condition rating. For the purposes of this example, assume the pitched roof is a CR3, the dormer roof is a Not Inspected (NI) and the bay roof has been categorised as a CR2 – because the lead flashings are loose but not resulting in a leak. The report will be ordered as follows (first to last):

- pitched roof (CR3) and taken forward to the summary in section C;
- flat roof to the bay (CR2), and then;
- the flat roof to the dormer (NI).

Each sub-element should be given its own sub-heading in bold and underlined for emphasis, together with its own condition rating. Although the reader will see that the roof covering has been allocated a CR3, on reading the full report the characteristics of the other roof should become evident.

Where there are several sub-elements and all have the same condition rating, you should identify the one which presents the greatest problem to the property owner and put this first. For example, if the main hipped roof, bay roof and flat roof to the rear single-storey extension are all CR2, the main roof should be listed and described first, because this is likely to be higher, larger and maybe more expensive to repair.

Cosmetic matters and minor defects

Cosmetic and minor items that do not impair the performance of a building element are not usually identified or reported, and the practice notes state '… disregard individual taste or fashion'. As all surveyors know, 'cosmetic items' can affect a buyer's decision and so affect the saleability. The element that includes the cosmetic issues should still be given a CR1 but may be emphasised in the report. For example, if the current owner has painted all rooms in the house bright orange, it will not affect the performance of the walls but it could discourage potential buyers. Your client must understand that if they want to protect the saleability of their new home, they will have to redecorate throughout.

The same goes for ordinary future maintenance, as this is outside the scope of the HBR. Nevertheless, where the design or materials employed in the construction of a building element are such that it needs to be maintained more frequently, or at a higher cost than might normally be expected, or if access is unusually complex, you should comment accordingly. The classic case is a flat roof covering, or the lime-based render on an older building. A CR1 will help the reader understand that compared to other similar properties, the element is in a satisfactory condition, but they also need to appreciate that the costs of maintaining that feature will be relatively high. I have seen many buyers who were only too keen to buy a period listed building, get upset when the estimates for repairs begin to drop on the door mat.

REPORT-WRITING EXAMPLES

I have already included a number of report-writing examples in chapter 5, which you may find useful. Below, I have included examples used during my training courses or included in previous sections of this book. To place them in context, I have used a number of photographs, but do not focus on the technical detail; assume that I assessed the element appropriately. I have included a commentary that 'deconstructs' the paragraph to show how and why I used the particular phrases for condition ratings 3, 2 and 1 respectively. I hope this is useful.

Stone built chimney stack

Figure 6.1(a) (b)

This is the descriptive section that outlines the element, how it is constructed, what is wrong with it, and the condition rating allocated.

The pots and stonework look loose and although it could have been in this state for years, the next strong wind could send part of it crashing through the ceiling. Therefore it is flagged as a safety hazard.

This sentence warns about the cost of repairing features at height. The repair could be small but the £500 to get up there may surprise the owner

The chimney stack to the main roof is made of stone with clay chimney pots. The stone stack is cracked, loose and is a safety hazard. **Condition rating 3**

This can be repaired, but it may be economic to rebuild the whole of the stack. To do this work safely, contractors will have to use appropriate access equipment (e.g. scaffolding, hydraulic platforms, etc.). This can increase the cost of the work.

You should check with your legal adviser before you do any work because the property is in a conservation area, and the owner of the neighbouring property may have a number of legal rights over this shared chimney (see section I).

It may look like a few loose stones on the ground but the chimney can be weaker than it looks. This sentence warns the client the extent of the work could increase.

These legal issues have been emphasised because they will have an impact on how the work is carried out. These are too important to tuck away in the legal section.

This is the advisory section that offers useful advice to the client and provides the 'added' value of the HBR.

Slate covered roof

This is the descriptive section. The technical term for the boarding has been used because for some the feature could be unusual.

Although the roof requires more than 'normal maintenance', I did not think it was serious or urgent.

This is the 'working at height' sentence. If there are a number of similar features, the sentence could be used once and other parts cross referenced to it.

E2 Roof covering
Sloping roof
The sloping roofs to the main house and rear addition are covered with slates fixed over timber boarding (sometimes called sarking boarding). Several slates are cracked and poorly secured. **Condition rating 2.**
You may find it economic to replace the whole roof covering rather than continue to repair. To do this work safely, contractors will have to use appropriate access equipment (e.g. scaffolding, hydraulic platforms, etc). This will increase the cost of the work.
You should check with your legal adviser before you do any work because the property is in a conservation area (see section I).

On a roof of this age, repair work could reveal corroded nail fixings, unseen deterioration, etc. This sentence advises that it might be a good time to recover the whole roof now.

Like the chimney before, it is important to emphasise the importance of this aspect.

This is the advisory section that offers useful advice to the client and provides the 'added' value of the HBR.

Gas boiler

Boiler C in Chapter 5 (please see figure 5.6 on page 55) received a CR1. This would be reported as follows:

H4 Heating

The home has gas central heating consisting of a wall-mounted condensing boiler, with radiators in every room. **Condition rating 1.**

Heating installations should be checked and serviced regularly (usually every year).

Sample report

The samples above, together with other report-writing examples elsewhere in this book should have given you an insight into how the comment boxes can be populated. To see how these can fit together, a sample report has been included in appendix 5.

COMPLETING SECTION I: ISSUES FOR YOUR LEGAL ADVISERS

The previous two chapters discussed the condition rating and reporting process, but there will be a number of issues that also have legal implications and these may influence a decision to purchase. Therefore you need to make sure section I interacts with the main body of the report. This is because the legal adviser will not read the report, but will simply look at section I for the matters they need to investigate. To make sure they have all the information they need, it may be appropriate to repeat parts of the main report as well as cross-referencing.

You should make any legal issues absolutely explicit to the client in the report. You should not simply tuck them away in section I. Instead, they should be described in the report and cross-referenced to section I. This will help the client see the implications of the legal issues in context, so when their adviser reports back, the client will know what to expect.

In other words, duplication between the main report and section I is acceptable – and indeed, necessary – to make sure 'the horse is taken to the water'. Hopefully, they may also drink.

Looking at this more practically, consider a few of the issues identified in the case studies discussed above (see figures 6.1(a) and (b); and 5.15, respectively).

The chimney stack (legal issues)

The legal issues of party walls and conservation areas were identified. The report advised: 'You should check with your legal adviser before any work is done because the property is in a conservation area, and the owner of the neighbouring property may have a number of legal rights over this shared chimney (see section I)'. This guidance could then be incorporated in section I in the following way:

> **I1 Regulation**
>
> **E1 Chimney stacks, E2 Roof Coverings, E3 Rain water pipes and gutters**
> The home is in a conservation area. You should ask your legal adviser to confirm this and explain the implications of owning a home in this type of area. Typical issues include having to get permission from the local council before you repair, alter or renovate the outside of the home, and having to use traditional building materials and experienced contractors. This can make the work costly.

> **I3 Other matters**
>
> **E1 Chimney stack**
> The chimney stack is shared with the neighbouring owner at 32 Acacia Avenue. You may need to get their agreement to the work before it can begin. You should ask your legal adviser to confirm this and explain the implications.

This may appear repetitive and long but I think it is worth the emphasis. These are important issues that both the legal adviser and the buyer need to be clear about. Surveyors can help by clearly signposting the matters. Where a property is in a conservation area, it may affect the repairs to a number of different elements and so there is a 'multiple' cross-reference.

The conservatory (legal issues)

The issue was the possible guarantees or warranty for the uPVC conservatory. The report included: 'You should ask your legal adviser if the conservatory has a guarantee or a warranty from the installer (see section I). If these do not exist, you should ask an appropriately qualified person to assess its quality.'

> **I2 Guarantees**
>
> You should ask your legal adviser to confirm whether the rear conservatory is covered by a guarantee or warranty and advise on the implications.

In many reports I have seen surveyors offer this advice, but what if a guarantee or a warranty does not exist? To avoid this confusion, I think we should add 'what if' guidance.

Guarantees and warranties

It is common to find that guarantees and warranties are lost, run out of time, are not worth the paper they are printed on, or were never issued in the first place. Without this documentation, the quality of the structure cannot be assured. Within the scope of the HBR, surveyors can make useful assessments of a conservatory but because the feature can be expensive to repair or to replace, and could have been an important factor in the decision to buy, further advice could be appropriate.

This advice is appropriate for guarantees and warranties. The most common difficulty arises over the remedial damp proof course, which reveals another common weakness in reports.

Where there are a number of drill holes on the outside of a wall and dampness problems inside, some surveyors hide behind a reference to the legal adviser suggesting the guarantee is checked. My advice is to be particularly careful here. In my experience, if a guarantee does exist, it is often restricted to specific parts of the property and so you should accurately identify the location of the problem. For example, where a chemical damp proof course has been injected but there is a dampness problem, section I could include the following note: 'You should ask your legal adviser to confirm whether the chemical damp proof course to the front and rear walls is covered by an enforceable guarantee. Because these walls are affected by dampness, you should ask them to advise on the implications.'

I picked up a good tip from a surveyor who suggested that 'enforceable' should be used because even where they do exist, most damp proofing guarantees are completely worthless. A useful 'what if' clause should be included in the affected element.

OVERALL OPINION

This section has been left until the end of this chapter, because it should be the last part of the HBR that you write.

Once you have inspected a property, reflected on all the pertinent issues, allocated the condition ratings, written the report and valued the property, you will then be in a position to give your overall opinion. In my opinion, this is what the HBR is all about.

The summary of condition ratings in section C will enable the client to quickly assess the condition of the property. The 'overall opinion' in the same section will help them take the '… most important decision of all: whether or not to proceed'. This section allows you to put the summary of the condition ratings in the context of the overall condition of the property.

The practice notes provide a good explanation of this process, so I want to consider how I arrived at the overall opinion for the Sample HBR in appendix 5.

Although it will be impossible to judge yourself, I felt that although there were some serious issues to resolve in this property, it certainly was not the worst of its kind. If the problems were addressed, '… I see no reason why there should be any special difficulty on resale in normal market conditions'. Looking at six standard paragraphs in the practice notes, 'Qualified YES 1' seems to be the closest. This reads:

Qualified YES 1 – Because defects identified are common in properties of this age and type
This property is considered to be a reasonable proposition for purchase at a price of £000,000, provided that you are prepared to accept the cost and inconvenience of dealing with the various repair/improvement works reported. These deficiencies are common in properties of this age and type. Provided that the necessary works are carried out to a satisfactory standard, I see no reason why there should be any special difficulty on resale in normal market conditions.

Reviewing the Summary of condition ratings, although it is a very crude measure, the relative proportions are:

CR3	10
CR2	7
CR1	5

With nearly half of the ratings at CR3 is this overall opinion too positive? Consider the following:

- There are structural concerns about the removed internal partition and the rear chimney breast. These will need to be resolved and will be costly, but they have not caused serious structural problems at the moment.
- There are two small areas of dampness that would be common in many homes of this type and age.
- The services pose the greatest potential problem. However, the heating and water heating are relatively new and the CR3 may be related to missing certificates rather than condition issues.

I find this review a little more reassuring. When using standard categories, to test my judgement I like to look at those above and below my choice. Looking at the one below (unqualified yes): it is clearly not appropriate. I would feel unable to give this opinion to a dwelling that has 10 CR3s:

Unqualified YES – Reasonable price/no condition rating 3s/no drawbacks
I am pleased to advise you that in my opinion this property is, on the whole, a reasonable proposition for purchase at a price of £000,000. I found no evidence of any significant problems, and I cannot foresee any special difficulties arising on resale in normal market conditions.

The other choice of standard paragraph would be Qualified Yes 2:

Qualified YES 2 – Because an element has a condition rating 3 or a risk identified in section J of the report
This property suffers from the exceptional disadvantage that [*state specific condition rating 3 or risk concisely*] and the agreed price of £000,000 reflects this. However, I must advise you that in my opinion this is likely to have an adverse effect on resale. I therefore urge you to consider with the utmost care whether you wish to proceed.

This is the opposite of the previous clause. Whoever buys this dwelling will have to spend time and money bringing it up to standard, but I doubt that it would have an adverse effect on the resale.

Consequently I settled for the Qualified YES 1. I did consider adjusting the wording to add a little more weight, but I thought this might be just splitting hairs. Taken together, the summary of the condition ratings and the overall opinion give the client a quick and complete impression of the property '… helping them to orientate themselves as they prepare to embark on the body of the report'.

7. Making the HomeBuyer Work

The previous chapters described the origins of the new Homebuyer Report, how to prepare and carry out the inspection, how to allocate condition ratings, advice on writing the report, and some aspects that affect market value.

Once you are confident with the new format and you have all the skills and knowledge to deliver it competently, there is nothing stopping you from launching your new service on the unsuspecting public. Well, the awful state of the market at the time of writing could possibly slow you down a bit ...

In this final chapter, I will briefly look at how you might exploit this new product and get ready for the future that promises to be forever changing.

EXPANDING THE MARKET SHARE

The old HSV did not go into the recession in good health.

- RICS estimates that only 20 per cent of the home buying public commission anything more than a mortgage valuation when buying a property.
- Of this 20 per cent, a proportion will be Building Surveys and other non-standard 'level two' products. Consequently, the HSV percentage could be as little as 15 per cent. In the words of Graham Ellis, the Professional Practice Manager of the RICS Residential Professional Group: 'There must be few sectors who would let 80 per cent of its customers walk away'.
- As suggested in chapter 1, the HSV itself has not been performing well. The format is looking tired and dated, and anecdotal information suggests that the reports are often poorly expressed, full of caveats and unnecessary waffle.

Even when the market picks up, will the level of our business return to its previous poor levels? Somehow I doubt it. The property market is changing. Lenders are more cautious, a high proportion of them are controlled by the government, and the public are becoming more cynical (or some would say 'more realistic') about the home buying process.

On my travels around the country I have met:

- those who have trained as domestic and commercial energy assessors and now operate in a very competitive market;
- those who have chosen to take dramatic cuts in salary or work part time; and
- one who has thrown in his damp meter and is now working as a postman!

Faced with these bleak prospects, what can the residential practitioner do? I am no business guru. After nine years in the public sector, 18 years in academia and only three years in the private sector, I am not sure I could help manage a business out of a wet paper bag, let alone help it face up to the economic challenge of a lifetime. However, what I can do is describe some of the initiatives others are taking and these may act as a catalyst for your own creative solutions.

Marketing the HBR

The launch of the new HomeBuyer Report gives an opportunity to market ourselves. We now have something new to say. It gives us a reason to knock on doors and explain this product, as well as the other services we can offer. It is a product that is modern, fresh and is focused on the needs of the buyer.

Many surveyors are already using this opportunity to promote their businesses:

- Syndicates of independent practitioners are organising breakfast meetings for key groups of clients including solicitors, agents and independent financial advisers. Using PowerPoint presentations, and RICS locally produced literature, they can make a positive impact on the key players in their area.
- Surveyors are establishing links with their local media in an effort to promote the public image of the chartered surveyor. This does (and could) include: writing topical columns in the local paper; taking part in phone-in advice shows on local radio stations; issuing regular press releases about a range of property related matters such as local auctions, new property regulations, impact of the budget, and so on.
- Some surveyors give talks to local organisations about property related matters. The Townswoman's Guild and the Women's Institute (WI) are just two examples. The WI have over 200,000 members nationally and nearly 1000 in my region alone. Most will be part or sole owners of a property.

- And how about linking up with a local community/advice centre to provide a 'drop-in' property advice session. Once organised, you could 'press release' the event and get good local media coverage as well.

There are other marketing techniques that we could all usefully exploit.

For example, you could produce your own email-based newsletter. Using standard software packages (Microsoft Word, Publisher, etc.), a digital camera and other software that can create PDF files, competent computer users can produce respectable publications that can put your company name on people's computer screens in a useful and positive way.

A typical edition of a four-page newsletter could include topical items such as: the launch of the new HBR; what to do about Japanese Knotweed; the pros and cons of 'buy to let'; and market prospects. If you keep it electronic it can be very cheap to produce and distribute, and can include good-quality colour photographs and diagrams.

If 'e-marketing' is going to work, you will need to build up a database of email addresses. Your existing clients are an obvious source, but many email addresses of commercial organisations can be found on the Internet in one form or another. A word of caution: the *Data Protection Act* 1998 aims to protect the privacy of individuals and restricts the collection or 'harvesting' of people's email addresses without their permission. For more information, you should visit the Information Commissioner website at www.ico.gov.uk

And finally, you could also go over to the 'dark side'. What I really mean is the Internet and I do not mean a website; I assume most practitioners have one of those already. Instead, I am talking about that mysterious world inhabited by people under the age of 30. According to my daughter, the following terms apparently mean something to millions of people:

MySpace

YouTube

Facebook

Twitter

At the time of writing, I am currently dipping my toe in this murky water so I am unable to give you any objective advice. However, try this for yourself. Type into your Internet search engine, www.youtube.co.uk and once you are on the site, type in 'Residential surveyors' in the search box. The small number of relevant videos will probably not surprise you. Carry out a new search for 'Home Inspections' and you will find hundreds of videos by American home inspector firms. These range from professionally produced short films, to amateur efforts by individual home inspectors. One larger practice has uploaded several small videos that discuss technical problems in houses and each one has five-figure viewing figures.

Did this activity lead to more business? Does YouTube enhance your practices image? I have no clear opinion. Yet I have a feeling there could be something to all this. For example, I recently visited a small but friendly garden centre and on a table in their café, I noticed a leaflet attached to the menu. It was headed 'Visit our Facebook site' where they promised topical tips, a discussion forum and lots more. In the event, the site was a little disappointing, but the 'topical tips' idea shows potential, and the 'forum' is a way of directly communicating with your customers. However, uncontrolled forums can cut both ways; watch out for the disgruntled customer who leaves abusive messages on your site.

The serious point here is that businesses are beginning to communicate with their customers through these social networking sites.

However, many cynics would suggest that Twitter and Facebook can become addictive. You can get distracted from 'real' work, especially if you gain a lot of friends or followers. Success involves developing a conversation with prospects and building a buzz round your business – not simply pumping out a load of 'buy me' messages. Small businesses that are aware of this are far more likely to succeed in this new interactive world.

Social networking sites could well be an exciting addition to your marketing armoury, but are probably best used in tandem with a number of other web activities

That is all the advice I have to give on the topic of social network sites, apart from an appeal for you to keep an open mind. If they can help an American president to get elected, can they help residential practitioners sell a few surveys? We must be ready for anything and if social networking sites can put us in contact with our clients then that is what we must embrace.

Support from RICS

To support the development of the HBR, RICS is planning to promote the product on a number of different levels:

- providing promotional material in an electronic format for members to use;
- raising awareness of the new HBR with other organisations including the National Federation of Property Professionals, the Law Society, FSA, CML, and so on;
- connecting consumers by working closely with Which?, specialist and general media, and the major property portals.

Your own efforts will be doubly effective if they run parallel with the RICS national campaigns.

BUILDING FOR THE FUTURE

During their HBR 'Roadshows', RICS representatives declared that the HBR is a 'declaration of intent' and is just the first step in the development of more RICS branded products. At the time of writing, two more initiatives are being planned:

- **The RICS Condition Report** – Although this has yet to be finalised, it promises to be the new RICS version of the Home Condition Report. In effect, it will be a stripped-down HBR keeping the condition ratings but omitting the advice and valuation. It should provide a more economic product for buyers and could also be used by sellers. RICS is currently lobbying the government to make this an authorised part of the HIP.
- **RICS Modular reports** – This is the new product that I am most excited by. These will be targeted on specific issues such as: a defect-specific report that investigates a matter of particular concern to the client; repair cost advice; flood prevention advice; an energy report (which could move beyond the EPC and place energy efficiency issues in context with property matters); and a safety hazard report, which could include a safety audit of the property, identify the hazards and make specific recommendations on how these could be reduced.

Modular reports would allow consumers to choose from a 'pick and mix menu' to suit their individual needs.

It is too early to predict whether developments of this type will be published, let alone successful, but they are a move in the right direction. What surveyors need is a range of services to offer our clients. Although RICS can help, much of the drive for change will come from surveyors themselves. My experience over the last few years has taught me that we must be flexible, adaptable and above all, self-sufficient. We cannot be reliant on the unpredictable policies of 'here today, gone tomorrow' politicians.

Although RICS is at last taking residential issues seriously, we cannot wait until their initiatives bear fruit. We have to find ways of developing our own relationships directly with our own clients. In doing so, we will be much less vulnerable than we have been in the past. To steal a few words from another famous American president (who did not have the advantage of YouTube), 'Ask not what RICS can do for you – ask what you can do for yourself.'

Appendices

Appendix 1
   (a) HBR site notes
   (b) HBR site notes (room-by-room supplement)
   (c) HBR site notes (for dictation)

Appendix 2 Sellers' questionnaire

Appendix 3 Market valuation analysis

Appendix 4 HBR site notes (sample)

Appendix 5 Completed report (sample)

APPENDIX 1 *(a) HBR site notes*

RICS HomeBuyer Report – Site notes

Surveyor's name		Company/address	
Date of inspection		Report reference	
Appointment arrangements and time of arrival		Time of departure	
Full address of the property Postcode:		Weather conditions	

Circumstances of inspection

Owner occupied / Tenanted / Vacant Comments:	Furnished / part furnished / empty / derelict Comments:
Access to HIP YES / NO Comments:	**EPC details** Name of DEA: Reference of certificate: Energy efficiency rating: Environmental impact rating:

Access to Property Information Questionnaire   YES / NO
Comments on content: (e.g. maintenance/safety checks of services, type of documentation checked; declared defects and alterations of property, etc.)

Type of property (please indicate)

House	Bungalow	Chalet	Flat	Maisonette	Shop/ commercial
Detached	Semi-detached	Terraced	End-terraced	Enclosed terr.	Other

Construction (broad indication of roof, walls, floor, etc.)

Date of construction/extension/conversion (describe evidence)

A Surveyor's Guide to the RICS HomeBuyer Service

Property address/Postcode: Page 2 of 26

Accommodation

Floor	Living rooms	Bedrooms	Bath/ shower	Separate toilet	Kitchen	Utility room	Conservatory	Other (name / details)
Lower ground								
Ground								
First								
Second								
Third								
Other								
Roof space								

Main services:	Drainage	Gas	Electricity	Water

General comments on services and other energy sources (wind turbine, PV, etc.):

Flats and maisonettes:

Which floor?		How many floors in block?	
No. of flats in block?		Converted?	
Purpose built?		Lift (describe)	
Commercial use? (describe, type, location, and % of block)			

Central heating:	Gas (mains)	Gas (LPG)	Electricity	Solid fuel
	Oil	Other (describe)		

Health and safety hazards, e.g.
- parking/road safety
- empty property
- access
- animals
- occupants
- asbestos fibres
- vermin, rats, pigeons, etc.
- voids/holes/unsafe areas
- tripping hazards
- ladders and fixed access equipment
- services – electrics, exposed wires, gas, LPG, plumbing, drains, etc.
- weather – high winds, etc.
- any other issues

Commentary (This section should give a short account of possible health and safety hazards in the property at the time of inspection. This should include an assessment of the level of risk (low; medium; high) and what action was taken to reduce the risk.)

Property address/Postcode:

Outside the property and the location

Checklist
- location of property
- neighbourhood
- adjoining uses
- local nuisances
- local planning applications
- parking/permits/access arrangements
- roads/rail/airports

Links to other sections
- *issues for legal adviser (section I)*
- *risks (section J)*

Please see section H for detailed comments on garages, outbuildings, grounds.

Facilities

Checklist
- shops
- schools
- transport – bus routes, railway station, etc.
- local services – doctors, etc.

Links to other sections
- *issues for legal adviser (section I)*
- *risks (section J)*

Please see section H for detailed comments on garages, outbuildings, grounds.

Local environment

Checklist
- flood risks
- radon
- EMFs
- mining activity
- contaminated/brownfield sites
- clay sub soils
- coastal erosion
- landslip area
- mundic area
- conservation area

Links to other sections
- *issues for legal adviser (section I)*
- *risks (section J)*

Please see section H for detailed comments on garages, outbuildings, grounds.

Property address/Postcode:

Site plan/External ....

Checklist
North point
Main outbuildings
Boundaries
Trees
Drain lines
Inspect chambers
Retaining walls

Property address/Postcode:

Floor plan ....

Checklist

North point

Dampness readings

Restrictions (furniture, floor cover, etc.)

Dimensions

Room heights

Property address/Postcode:

Floor plan ....

Checklist

North point

Dampness readings

Restrictions (furniture, floor cover, etc.)

Dimensions

Room heights

Property address/Postcode:

Section E – Outside the property	
E1 Chimney stack	**Description of construction and condition**
Checklist • description, number and location • flue terminals • chimney pots • flaunching • condition of stacks (leaning, cracked, sulfate attack, etc.) • pointing, render, other finishes • aerials and satellite dishes • flashings and soakers at the junction with the roof covering (but not including the roof covering) • any party wall issues arising from chimney condition • listed building/conservation areas issues **Links to other sections** • *cross reference to other affected elements* • *issues for legal adviser (section I)* • *risks (section J)*	
E2 Roof coverings	**Description of construction and condition**

Pitched roof
• type of covering and general condition including:
  – *the slope/pitch of the roof*
  – *fixing of tiles, slates*
  – *temp. repairs*
• roofing felt/sarking boarding – presence, type, condition
• ridges and hip tiles
• verge and eaves details
• open valleys
• valley gutters ('butterfly' roofs), parapets gutters, their linings and outlets
• lead flashings to up-stands, dormers, etc. (*but not the flashings to the chimney*)
• roof lights/roof windows
• dormers (*usually including the flat or pitched over the feature*)
• any party wall issues arising from the roof condition
• listed building/conservation area issues
 	
Links to other sections
• *cross reference to other affected elements*
• *issues for legal adviser (section I)*
• *risks (section J)*	

Property address/Postcode:

E2 Roof coverings (cont.)	Description of construction and condition

Checklist
Flat roof
• type of covering and condition including:
  – *the falls on the roof*
  – *pooling of water*
  – *temp. repairs*
  – *solar protection*
• lead flashings to up-stands (*but not any chimney flashings*)
• other penetrations (*e.g. SVP, vertical flues, etc.*)
• verge and eaves details
• any integral gutters
• roof lights and upstands
• any party wall issues arising from the roof condition;
• listed building/conservation areas
 	
Links to other sections
• *cross reference to other affected elements*
• *issues for legal adviser (section I)*
• *risks (section J)*	

E3 Rainwater pipes and gutters	**Description of construction and condition**

Checklist
• type of gutters, gutter supports and condition including:
  – *falls to gutter*
  – *position and adequacy of downpipes, hopper, etc.*
  – *debris/blockage*
  – *leaks and damage to other parts of building*
• swan necks and off-sets
• rainwater downpipes
• hopper heads
• rainwater shoes, etc.
• listed building/conservation areas;
• any party issues with gutters and downpipes, e.g. shared roof drainage, etc.
 	
Links to other sections
• *cross reference to other affected elements*
• *issues for legal adviser (section I)*
• *risks (section J)*	

Property address/Postcode:

E4 Main walls	Description of construction and condition

Checklist
• type, thickness and general condition of the wall(s) including:
  – *wall ties or sulfate attack*
  – *wall tie replacement*
  – *pointing*
  – *rendering/external finishes*
  – *cracking/distortion*
  – *bulging/verticality*
  – *efflorescence*
• cladding:
  – *vertical tiles;*
  – *PVC*
  – *timber*
• damp proof course – type, position, and condition
• evidence of injection holes
• relationship with external ground levels
• presence of dampness to the inside face of the outside wall from rising and penetrating dampness and condensation
• the type and condition of the lintels, jambs and sills
• parapet walls and copings including party up-stand walls
• timber frame walls
• type and extent of insulation to the walls
• any party wall issues arising from the wall condition
• any listed building/conservation area issues
 	
Links to other sections
• *cross reference to other affected elements*
• *issues for legal adviser (section I)*
• *risks (section J)*	

Property address/Postcode:

E5 Windows	Description of construction and condition
Checklist • the type and general condition of the windows • orientation of windows • opening of window sample • single/double/secondary glazing • seals to frame • damp penetration around frames • draughts/trickle vents • seals to double glazed units • replacement window issues • escape in the case of fire • low level glass/safety glass • security • listed building/conservation area issues **Links to other sections** • *cross reference to other affected elements* • *issues for legal adviser (section I)* • *risks (section J)*	
E6 Outside doors (and patio doors)	**Description of construction and condition**
Checklist • type/condition of door and frame • level threshold? If level access, exclusion of water, etc. • glazing and safety glass issues, safety glass, etc. • security • draughts/damp penetration around frame • seals to double glazed units • listed building/conservation area issues **Links to other sections** • *cross reference to other affected elements* • *issues for legal adviser (section I)* • *risks (section J)*	

Property address/Postcode:

E7 Conservatories and porches	Description of construction and condition
Checklist • type and condition • inappropriate use as habitable room • single/double glazing • safety - structural stability and safety glazing • snow/debris guards to edge of dwelling roof • separation from rooms of the main dwelling • waterproofing at junction with dwelling • opening of sample of windows • security of windows/doors • effect on escape in the case of fire from the upper windows of the dwelling • guarantee/warranty • building regulation and/or planning approval • listed building/conservation area issues **Links to other sections** • *cross reference to other affected elements* • *issues for legal adviser (section I)* • *risks (section J)*	
E8 Other joinery and finishes	**Description of construction and condition**

Checklist
• type and condition of joinery including:
  – *signs of rot*
  – *state of the decorations*
• fascias, soffits and eaves details generally
• bargeboards and verge details;
• ornamental joinery (finials, mock Tudor panelling, etc.);
• timber and PVC cladding fixed to main walling
• possible asbestos containing materials
 	
Links to other sections
• *cross reference to other affected elements*
• *issues for legal adviser (section I)*
• *risks (section J)*	

Property address/Postcode:

E9 Other	Description of construction and condition
Checklist • roof terraces – useable flat roof areas over other accommodation below • balconies • large bay windows where distinctive and different in nature to the construction of the main walls • external stairways and fire escapes, etc. For each feature, include all aspects of the feature, e.g.: • structural aspects – fixings, support, stability, etc. • the walls, floor and roof (if present) to the feature • railings, handrails, etc. **Links to other sections** • *cross reference to other affected elements* • *issues for legal adviser (section I)* • *risks (section J)*	

Other comments
Include any other external matters here.

Property address/Postcode:

Section F – Inside the property

Restrictions on inspection

Include details of any restrictions on internal inspection:

F1 Roof structure and roof space	Description of construction and condition
Checklist **Pitched roofs** • description of design, construction and condition of the roof structure • sample moisture readings • alteration of the roof structure (full attic conversions can be included under F9 Other) • roof strengthening • lateral restraint, spreading of roof structure • the use and misuse of the roof space • wood rot and wood boring beetle • party walls – lack of fire breaks, gaps between dwellings • flooring in loft • possible asbestos containing materials • access into loft • ventilation in roof space • level of insulation • Birds, bats, vermin, insects, etc. **Flat roof structure** • access/restrictions to flat roof structure • sample moisture meter readings • ventilation to flat roof void • wood rot and wood boring insect (Services in the roof space to be included under appropriate elements in section G) **Links to other sections** • *cross reference to other affected elements* • *issues for legal adviser (section I)* • *risks (section J)*	

Property address/Postcode:

F2 Ceilings	Description of construction and condition
Checklist • description of type and condition including cracking • safety issues – stability of ceiling, bulging, etc. • cornices/centrepieces other decorative features • finishes, including decorations (heavy lining paper, textured coating, etc.) • possible asbestos containing materials, textured decorative coatings, asbestos board, etc. • Listed building features and issues **Links to other sections** • *cross reference to other affected elements* • *issues for legal adviser (section I)* • *risks (section J)*	

F3 Walls and ceilings	Description of construction and condition
Checklist • description of type and condition • loading or non-load-bearing partitions • structural movement, cracking, distortion, binding doors, etc. • dampness to internal walls including condensation mould growth • non-robust partitions, poor sound insulation, low level of fire resistance, etc. • possible asbestos containing material, asbestos board, textured coating, etc. • finishes including decorations • wall tiling (including splash back wall tiling) • dado panelling • structural alterations – through lounge, new door openings, level of support, etc **Links to other sections** • *cross reference to other affected elements* • *issues for legal adviser (section I)* • *risks (section J)*	

Property address/Postcode:

F4 Floors	Description of construction and condition

Checklist
• description of type and condition (floor by floor)
• solid floor:
  – *cracking*
  – *settlement*
  – *swelling*
• suspended timber floor:
  – *stamp test and vibration*
  – *access hatch?*
  – *ventilation to sub floor void*
• description of floor finishes and their condition
• dampness to all floor types
• wood boring insects
• wood rot
 	
Links to other sections
• *cross reference to other affected elements*
• *issues for legal adviser (section I)*
• *risks (section J)*	

F5 Fireplaces, chimney breasts and flues	Description of construction and condition

Checklist
Description of fireplaces, chimney breasts and flues and their condition
• **open, solid fuel fireplaces**:
  – *flue/chimney maintenance sweeping*
  – *lining*
  – *Part J approvals*
  – *noncombust. hearth*
  – *certificates, invoices, etc.*
  – *fixed external ventilation*
  – *hearth*
• **chimney breasts**:
  – *open or blocked off*
  – *if blocked – ventilation to flue*
• terminal at top of flue at chimney level
• condition of chimney breast in roof space:
  – *structural alterations unsupported chimney breast*
  – *dampness to base and face of chimney breasts*
	Please note: Actual gas and oil heaters/fires to be included under G4: Heating. This element includes the chimney breast from the lowest level in the house, through all rooms, through any roof space and finishes at the underside of the roof covering. The element changes to 'chimney stack' (E1) above the roof covering where it becomes an external element.

Property address/Postcode: Page 16 of 26

F5 Fireplaces, etc. (cont.)	Description of construction and condition

• independent open vertical flues:
  – description and fixing type, and condition
  – suitability of terminal position
• proximity of flue to combustible materials, etc.
• possible asbestos materials
• listed building issues
 	
Links to other sections
• *cross reference to other affected elements*
• *issues for legal adviser (section I)*
• *risks (section J)*
	Please note: Balanced/fanned flues to be included under G4 Heating. Include here vertical open flues not contained within a chimney.

F6 Built-in fittings	Description of construction and condition

Checklist
Description, adequacy and general condition
 	
• **kitchen units, cupboards and worktops:**
  – *the waterproof sealant but not the splash back tiling (see F3)*
• **kitchen sinks:**
  – *the taps (but not the water pipes to them – G3 Water)*
  – *the slotted wastes (but not the trap or the waste pipe: G6 Drainage)*
• dampness affecting the features described above
• wood-boring insects affecting the features described above
• operation and functionality – operable drawers, damaged hinges, etc.
• age and modernity – appeal to purchasers
 	
Links to other sections
• *cross reference to other affected elements*
• *issues for legal adviser (section I)*
• *risks (section J)*
	(Built-in kitchen and other fittings, not including the appliances.)

Other comments

Property address/Postcode:

F7 Woodwork	Description of construction and condition
Checklist Description and general condition: • internal doors, frames/linings, architraves, etc. • internal parts of windows – jamb linings, sill boards, architraves, shutters, etc. • built in cupboards – for example within chimney alcoves, meter cupboards, understairs • skirtings, dado rails, picture rails • staircases – treads and risers, strings, balusters, handrails, newel posts: – *safety hazards – missing balusters, broken treads, etc.* • finishes to features above including decorations • dampness affecting the features described above • wood rot and wood-boring beetle affecting the features described above **Links to other sections** • *cross reference to other affected elements* • *issues for legal adviser (section I)* • *risks (section J)*	

F8 Bathroom fittings	Description of construction and condition
Checklist Description, adequacy and general condition: • baths, bidets, wash hand basins, their taps and slotted wastes but **not** the water pipes to them (G3 and G5), and **not** the traps and waste pipe (G6) • shower trays, shower cubicles but **not** the mixer valve (G3) or electric shower (G1) • WCs, their cisterns and overflows. To include junction between the WC outgo and the drainage pipe but not the drainage pipe itself (G6) • bath panels and other built-in fittings associated with the sanitary appliances • the sealant between the appliance and the adjacent surfaces but not the splash back tiles (see F3) • leaks and dampness caused by the bathroom fittings • wood rot and wood-boring insects affecting the features described above **Links to other sections** • *cross reference to other affected elements* • *issues for legal adviser (section I)* • *risks (section J)*	

Property address/Postcode:

F9 Other	Description of construction and condition
Checklist This section is for anything internal that does not sit comfortably in earlier sections in section F. This could typically include: • **Roof space conversion** – include here where a total conversion has been carried out. To give a complete appraisal of the conversion, include all elements here. For example: – *structural aspects – alteration of roof, strengthening of ceilings and other floors* – *access to room – adequacy and safety of staircase, etc.* – *thermal insulation – including ventilation to roof spaces* – *fire escape route – fire resistance of partitions, ceilings, doors, etc.* – *regulation and permissions – building regulation, planning permission, listed building/conservation, etc.* • **Cellar/basements** – where the cellar/basements are used for habitable purposes, to give a complete appraisal, include all elements here. For example: – *structural aspects – removal of partitions, strengthening of ceilings and floors* – *access to room – adequacy and safety of staircase, etc.* – *fire escape and fire safety – additional fire resistance to ceiling, inner room issues, etc.* – *daylight and ventilation to rooms and spaces – compare with building regs* – *dampness – tanking issues, guarantees, etc.* • **Flats** – to include those internal common parts of a block: – *internal hallways, landings and stairs* – *doors and windows* – *built-in cupboards, meter cupboards, etc.* – *asbestos containing materials* – *asbestos management plan* **Links to other sections** • *cross reference to other affected elements* • *issues for legal adviser (section I)* • *risks (section J)*	

Property address/Postcode:

Section G - services	
Restrictions on inspection	
Include details of any restrictions on the inspection of the services:	
G1 Electricity	**Description of construction and condition**
Checklist • description and general condition • evidence of installation, alteration or maintenance certification (e.g. Part P approval, minor works cert., periodic inspections and reports, etc. **Main issues** (broad indicators only): • is there a mains supply • on-peak/off-peak supply • location of the meter and consumer unit/fuse board • separate from gas meter • is there a RCD/MCB • equipotential bonding in the usual places • type and condition of visible wiring • condition of a sample of the range of light fittings and switch gear • type and condition of fixed electrical appliances including heaters, storage radiators, electric showers, instant water heaters, etc. (see also G5 water heating) • bath and shower rooms – nature and suitability of electrical fittings • external installations – garages, outbuildings, external sockets, garden lighting, water feature pumps, etc. • external installations – covered by RCD **Links to other sections** • *cross reference to other affected elements* • *issues for legal adviser (section I)* • *risks (section J)*	The following general advice will be printed in every report: Safety warning: Periodic inspection and testing of electrical installations is important to protect your home from damage and to ensure the safety of the occupants. Guidance published by the Institution of Electrical Engineers recommends that inspections and testing are undertaken at least every ten years and on change of occupancy. All electrical installation work undertaken after 1 January 2005 should be identified by an Electrical Installation Certificate.

A Surveyor's Guide to the RICS HomeBuyer Service

Property address/Postcode:

G2 Gas/Oil	Description of construction and condition

Please note: this section covers the gas and oil supply pipework from the utility company's supply (or the storage vessel) through to the heating appliance itself. All matters relating to the appliance are to be reported under G4 Heating.
 	
Checklist
• description and general condition
• evidence of installation, alteration or maintenance certification (e.g. Gas Safe Registered engineer reports, OFTEC certs., service agreement, etc.)
• **mains gas installations:**
• does it have a mains supply and is it connected
• meter location and condition – is it positioned on an escape route
• the route, nature and condition of the pipework – well fixed and supported
• have there been alterations and are they properly certified
• **LPG installations:**
• are the storage cylinders in a suitable location:
  – *boundaries*
  – *proximity to buildings (especially basements), other storage tanks*
• condition of tank, covered with foliage, etc
• the route, nature and condition of the pipework
• have there been alterations, and are they properly certified?
• **oil installations:**
• is this satisfactory, including:
• the position of tank:
  – *proximity to dwelling, boundaries, watercourses, drains, etc.*
  – *seen from filling point*
• nature of the tank:
  – *internal or external bunded*
  – *condition of tank*
 	
Links to other sections
• *cross reference to other affected elements*
• *issues for legal adviser (section I)*
• *risks (section J)*
	The following general advice will be printed in every report:
	Safety Warning: Regular inspection, testing, maintenance and servicing of all gas/oil appliances and equipment should be undertaken by a registered 'competent person' and in accordance with the manufacturer's instructions. This is important to ensure that such equipment is working correctly to minimise the risk of fire and carbon monoxide poisoning as well as leakages of carbon dioxide and other greenhouse gases to the atmosphere. For further advice contact Gas Safe Register for gas installations, and OFTEC for oil installations.

Other comments

Property address/Postcode:

G3 Water	Description of construction and condition

Checklist
General description of the system and condition including:
• the piped water supply from the boundary of the property:
  – *pipe material (lead?)*
  – *external stop valve*
  – *depth of pipe?*
  – *water meter*
  – *where does it enter building?*
• Internally, where the rising main appears:
  – *stop valve – accessibility*
  – *insulation (especially underfloor and in roof space)*
  – *material – lead? dissimilar materials?*
  – *support to pipework*
• cold water storage and expansion tanks:
  – *support to tank*
  – *tank type and condition*
  – *lid, thermal insulation*
  – *overflow*
• backflow prevention to:
  – *bidet, etc.*
  – *garden hose pipe,*
  – *irrigation system*
 	
Private water supplies
Main issues to include
• description and general condition
• private supply only or combined with mains
• nature of source
• has supply been checked and certified as wholesome source of supply
• shared supply with neighbours
 	
Links to other sections
• *cross reference to other affected elements*
• *issues for legal adviser (section I)*
• *risks (section J)*
	Please note: This element covers cold water supply service to combination boilers, water heaters, hot water cylinders but not the pipe work beyond them (see G5). It covers cold water storage tanks and the down service beyond, but **not** the taps of appliances.

Other comments

Property address/Postcode:

G4 Heating	Description of construction and condition

Checklist
General description and condition of the heating system(s):
• evidence of certification for installation, maintenance, repair etc. (see also G2)
• boilers (including combination boilers that also provide hot water)
• gas and oil fires and heaters
• electric heating (but not the electricity supply to them)
• heat pumps
• other forms (describe)
• heat distribution pipework
• heat emitters – radiators, panel heaters, under-floor heating pipework, etc.
• ancillary equipment including:
  – *heating controls*
  – *expansion vessels and safety controls*
• ventilation issues for heating appliances (where appropriate)
Please note: free standing solid fuel heating sources (incl. Aga) should be covered here. Where in fireplace include in F5 but distribution pipework, etc. covered under this element.
 	
Links to other sections
• *cross reference to other affected elements*
• *issues for legal adviser (section I)*
• *risks (section J)*
	Please note: Most heating appliances will be subject to regulations governing installation of gas, oil, solid fuel and electric appliances previously described. Cross reference to appropriate section where certification is required.

G5 Water heating	Description of construction and condition
Checklist Include the hot water heating appliance where it provides hot water only, for example: • gas water heater • multi-point water heater • electric under-counter heater • instantaneous electric water heater; • solar water/PV heating Other parts of hot water system to include: • the hot water cylinder • insulation to tank and primary pipe work • expansion pipe • immersion heaters • cylinder thermostats • all distribution pipework between cylinder and the appliances (but not the taps) • the distribution pipework between water heating source and the various appliances **Links to other sections** • *issues for legal adviser (section I)* • *risks (section J)*	**Please note:** Where hot water provided by combination heating, hot water boilers, heat pumps, etc. include under G4. Evidence of certification if not covered in G2/G4.

Property address/Postcode:

G6 Drainage	Description of construction and condition

Checklist
General description and condition of drainage system.
Above ground drainage:
• traps and wastes from sinks, washing machines, dishwashers, baths, bidets, WCs, showers, wash hand basin, etc.
• pipe material, condition, leaks
• length of wastes, cleaning access
• slope of branch connections
• soil and vent pipes, two pipe systems, etc.
• proximity of stack to windows
• stub stacks
• air admittance valves, etc.
Below ground drainage:
• gullies – type, condition, grid, blocked
• clearing access to drains
• inspection chambers:
  – *position*
  – *cover/condition*
  – *walls/benching/channel*
  – *standard of connections*
  – *Bldg regs for new work*
  – *interceptor*
  – *blocked*
• underground drains:
  – *type*
  – *building over drains*
  – *trees close by?*
  – *depth and approx fall?*
• design of system:
  – *separate soil and surface*
  – *combined system*
  – *inappropriate connections*
  – *join with other properties – private sewer?*
Private drainage including:
• design and type:
  – *cess pit*
  – *septic tank (two chamber)*
  – *small sewerage treatment system (e.g. Klargester, etc)*
• Other types:
  – *soakaway*
  – *reed beds*
  – *composting toilets.*
• Evidence of maintenance, permission to discharge, etc.
• Shared system
• Approx. size? Adequacy
• Evidence of misuse
  – *RWP connection*
  – *Extra WCs to small tank*
 	
Links to other sections
• *cross reference to other affected elements*
• *issues for legal adviser (section I)*
• *risks (section J)*	

Property address/Postcode:

G7 Common services	Description of construction and condition
Checklist General description and condition of the facilities to the block General comments in relation to: • refuse chutes, bin stores, etc. • lifts – passenger and goods • entry-phone systems • CCTV • common areas – heating and lighting • communal heat and power • renewable energy systems **Links to other sections** • *cross reference to other affected elements* • *issues for legal adviser (section I)* • *risks (section J)*	

Other comments

Property address/Postcode:

Section H – Grounds (including shared areas for flats)

Restrictions on inspection

Include details of any restrictions on the inspection of the services:

H1 Garage	Description of construction and condition

Checklist
General description, location, and condition including all elements and services, including:
• functionality:
  – *size for modern cars*
  – *access to entrance*
• condition:
  – *structure*
  – *walls/cladding*
  – *floor (inspection pits?)*
  – *roof – leaks, fragile covering*
  – *asbestos containing materials*
  – *windows/doors*
• services:
  – *electrics (RCD protection?)*
  – *heating*
  – *water*
  – *drainage*
• unauthorised use – commercial use, office, habitable, etc.
• conservation area, listed buildings issues
Please note: Integral garages should be including in the main report (mostly section F) under the elemental rating system.
 	
Links to other sections
• *issues for legal adviser (section I)*
• *risks (section J)*	

H2 Other	Description of construction and condition

Checklist
Other permanent outbuildings:
• summer houses
• permanent greenhouses that are to remain
• follies
• substantial garden sheds that are to remain
• leisure buildings (but not the leisure facilities inside)
• for each of the above:
  – *general description and condition including roof, walls, floors, windows, doors, etc.*
  – *services*
  – *electrics (RCD protection?)*
  – *heating*
  – *water*
  – *drainage*
• unauthorised use – commercial use, office, habitable, etc.
• conservation area, listed building issues
 	
Links to other sections
• *issues for legal adviser (section I)*
• *risks (section J)*	

Property address/Postcode:

H3 General	Description of construction and condition
Checklist To include a general description of the grounds and/or site including: • boundaries – condition and likely ownership, changed positions, etc. • retaining walls – construction, type, condition, etc. • terraces and patios • conservation and listed building issues • gardens – general description, approx. size, landscape/garden design features • drives and paths – type and condition • steps and changes of level – type, construction safety • trees – type, location to buildings and drains, maintenance, safety • ponds, lakes, reed beds, drainage fields, water courses, etc. – type, size, condition, ownership implications, safety • rights of way/easements across site/wayleaves, etc. • wider location/environment issues where impinge within boundaries, e.g. contamination, HV pylons, HV sub stations, landslip, etc. **Links to other sections** • *issues for legal adviser (section I)* • *risks (section J)*	

Other comments

APPENDIX 1 *(b) HBR site notes (room-by-room supplement)*

RICS HomeBuyer Report – Site notes (room-by-room supplement)

Surveyor's name		Company/address	
Date of inspection		Report reference	
Appointment arrangements and time of arrival		Time of departure	
Full address of the property Postcode:		Weather conditions	

Circumstances of inspection	
Owner occupied / Tenanted / Vacant Comments:	Furnished / part furnished / empty / derelict Comments:

Access to HIP YES / NO Comments:	EPC details Name of DEA: Reference of certificate: Energy efficiency rating: Environmental impact rating:

Access to Property Information Questionnaire   YES / NO
Comments on content: (e.g. maintenance/safety checks of services, type of documentation checked; declared defects and alterations of property, etc.)

Type of property (please indicate)

House	Bungalow	Chalet	Flat	Maisonette	Shop/ commercial
Detached	Semi-detached	Terraced	End-terraced	Enclosed terr.	Other

Construction (broad indication of roof, walls, floor, etc.)

Date of construction/extension/conversion (describe evidence)

A Surveyor's Guide to the RICS HomeBuyer Service

Property address/Postcode:

Accommodation

Floor	Living rooms	Bedrooms	Bath/ shower	Separate toilet	Kitchen	Utility room	Conservatory	Other (name / details)
Lower ground								
Ground								
First								
Second								
Third								
Other								
Roof space								

Main services:	Drainage	Gas	Electricity	Water
General comments on services and other energy sources (wind turbine, PV, etc.):				

Flats and maisonettes:			
Which floor?		How many floors in block?	
No. of flats in block?		Converted?	
Purpose built?		Lift (describe)	
Commercial use? (describe, type, location, and % of block)			

Central heating:	Gas (mains)	Gas (LPG)	Electricity	Solid fuel
	Oil	Other (describe)		

Health and safety hazards, e.g.	**Commentary** (This section should give a short account of possible health and safety hazards in the property at the time of inspection. This should include an assessment of the level of risk (low; medium; high) and what action was taken to reduce the risk.)
<ul><li>parking/road safety</li><li>empty property</li><li>access</li><li>animals</li><li>occupants</li><li>asbestos fibres</li><li>vermin, rats, pigeons, etc.</li><li>voids/holes/unsafe areas</li><li>tripping hazards</li><li>ladders and fixed access equipment</li><li>services – electrics, exposed wires, gas, LPG, plumbing, drains, etc.</li><li>weather – high winds, etc.</li><li>any other issues</li></ul>	

Property address/Postcode:

Outside the property and the location

Checklist
- location of property
- neighbourhood
- adjoining uses
- local nuisances
- local planning applications
- parking/permits/access arrangements
- roads/rail/airports

Links to other sections
- *issues for legal adviser (section I)*
- *risks (section J)*

Please see section H for detailed comments on garages, outbuildings, grounds.

Facilities

Checklist
- shops
- schools
- transport – bus routes, railway station, etc.
- local services – doctors, etc.

Links to other sections
- *issues for legal adviser (section I)*
- *risks (section J)*

Please see section H for detailed comments on garages, outbuildings, grounds.

Local environment

Checklist
- flood risks
- radon
- EMFs
- mining activity
- contaminated/brownfield sites
- clay sub soils
- coastal erosion
- landslip area
- mundic area
- conservation area

Links to other sections
- *issues for legal adviser (section I)*
- *risks (section J)*

Please see section H for detailed comments on garages, outbuildings, grounds.

Property address/Postcode:

Site plan/External ....

Checklist
North point
Main outbuildings
Boundaries
Trees
Drain lines
Inspect chambers
Retaining walls

Property address/Postcode:

Floor plan ....

Checklist

North point

Dampness readings

Restrictions (furniture, floor cover, etc.)

Dimensions

Room heights

Property address/Postcode:

Floor plan ....

Checklist

North point

Dampness readings

Restrictions (furniture, floor cover, etc.)

Dimensions

Room heights

Property address/Postcode:

Section E – Outside the property	
E1 Chimney stack	**Description of construction and condition**
Checklist • description, number and location • flue terminals • chimney pots • flaunching • condition of stacks (leaning, cracked, sulfate attack, etc.) • pointing, render, other finishes • aerials and satellite dishes • flashings and soakers at the junction with the roof covering (but not including the roof covering) • any party wall issues arising from chimney condition • listed building/conservation area issues **Links to other sections** • *cross reference to other affected elements* • *issues for legal adviser (section I)* • *risks (section J)*	
E2 Roof coverings	**Description of construction and condition**

Pitched roof
• type of covering and general condition including:
  – *the slope/pitch of the roof*
  – *fixing of tiles, slates*
  – *temp. repairs*
• roofing felt/sarking boarding – presence, type, condition
• ridges and hip tiles
• verge and eaves details
• open valleys
• valley gutters ('butterfly' roofs), parapets gutters, their linings and outlets
• lead flashings to up-stands, dormers, etc. (*but not the flashings to the chimney*)
• roof lights/roof windows
• dormers (*usually including the flat or pitched over the feature*)
• any party wall issues arising from the roof condition
• listed building/conservation area issues
 	
Links to other sections
• *cross reference to other affected elements*
• *issues for legal adviser (section I)*
• *risks (section J)*	

Property address/Postcode: Page 8 of 25

E2 Roof coverings (cont.)	Description of construction and condition

Checklist
Flat roof
• type of covering and condition including:
   – *the falls on the roof*
   – *pooling of water*
   – *temp. repairs*
   – *solar protection*
• lead flashings to up-stands (*but not any chimney flashings*)
• other penetrations (*e.g. SVP, vertical flues, etc.*)
• verge and eaves details
• any integral gutters
• roof lights and upstands
• any party wall issues arising from the roof condition;
• listed building/conservation areas
 	
Links to other sections
• *cross reference to other affected elements*
• *issues for legal adviser (section I)*
• *risks (section J)*	

E3 Rainwater pipes and gutters	Description of construction and condition

Checklist
• type of gutters, gutter supports and condition including:
   – *falls to gutter*
   – *position and adequacy of downpipes, hopper, etc.*
   – *debris/blockage*
   – *leaks and damage to other parts of building*
• swan necks and off-sets
• rainwater downpipes
• hopper heads
• rainwater shoes, etc.
• listed building/conservation areas;
• any party issues with gutters and downpipes, e.g. shared roof drainage, etc.
 	
Links to other sections
• *cross reference to other affected elements*
• *issues for legal adviser (section I)*
• *risks (section J)*	

Property address/Postcode:

E4 Main walls	Description of construction and condition

Checklist
• type, thickness and general condition of the wall(s) including:
  – *wall ties or sulfate attack*
  – *wall tie replacement*
  – *pointing*
  – *rendering/external finishes*
  – *cracking/distortion*
  – *bulging/verticality*
  – *efflorescence*
• cladding:
  – *vertical tiles;*
  – *PVC*
  – *timber*
• damp proof course – type, position, and condition
• evidence of injection holes
• relationship with external ground levels
• presence of dampness to the inside face of the outside wall from rising and penetrating dampness and condensation
• the type and condition of the lintels, jambs and sills
• parapet walls and copings including party up-stand walls
• timber frame walls
• type and extent of insulation to the walls
• any party wall issues arising from the wall condition
• any listed building/conservation areas issues
 	
Links to other sections
• *cross reference to other affected elements*
• *issues for legal adviser (section I)*
• *risks (section J)*	

Property address/Postcode:

E5 Windows	Description of construction and condition
Checklist • the type and general condition of the windows • orientation of windows • opening of window sample • single/double/secondary glazing • seals to frame • damp penetration around frames • draughts/trickle vents • seals to double glazed units • replacement window issues • escape in the case of fire • low level glass/safety glass • security • listed building/conservation area issues **Links to other sections** • *cross reference to other affected elements* • *issues for legal adviser (section I)* • *risks (section J)*	
E6 Outside doors (and patio doors)	**Description of construction and condition**
Checklist • type/condition of door and frame • level threshold? If level access, exclusion of water, etc. • glazing and safety glass issues, safety glass, etc. • security • draughts/damp penetration around frame • seals to double glazed units • listed building/conservation area issues **Links to other sections** • *cross reference to other affected elements* • *issues for legal adviser (section I)* • *risks (section J)*	

Property address/Postcode:

E7 Conservatories and porches	Description of construction and condition
Checklist • type and condition • inappropriate use as habitable room • single/double glazing • safety - structural stability and safety glazing • snow/debris guards to edge of dwelling roof • separation from rooms of the main dwelling • waterproofing at junction with dwelling • opening of sample of windows • security of windows/doors • effect on escape in the case of fire from the upper windows of the dwelling • guarantee/warranty • building regulation and/or planning approval • listed building/conservation area issues **Links to other sections** • *cross reference to other affected elements* • *issues for legal adviser (section I)* • *risks (section J)*	

E8 Other joinery and finishes	Description of construction and condition

Checklist
• type and condition of joinery including:
  – *signs of rot*
  – *state of the decorations*
• fascias, soffits and eaves details generally
• bargeboards and verge details;
• ornamental joinery (finials, mock Tudor panelling, etc.);
• timber and PVC cladding fixed to main walling
• possible asbestos containing materials
 	
Links to other sections
• *cross reference to other affected elements*
• *issues for legal adviser (section I)*
• *risks (section J)*	

Property address/Postcode:

E9 Other	Description of construction and condition
Checklist • roof terraces – useable flat roof areas over other accommodation below • balconies • large bay windows where distinctive and different in nature to the construction of the main walls • external stairways and fire escapes, etc. For each feature, include all aspects of the feature, e.g.: • structural aspects – fixings, support, stability, etc. • the walls, floor and roof (if present) to the feature • railings, handrails, etc. **Links to other sections** • *cross reference to other affected elements* • *issues for legal adviser (section I)* • *risks (section J)*	

Other comments
Include any other external matters here.

Property address/Postcode:

Section F – Inside the property

Restrictions on inspection
Include details of any restrictions on internal inspection:

F1 Roof structure and roof space	Description of construction and condition
Checklist **Pitched roofs** • description of design, construction and condition of the roof structure • sample moisture readings • alteration of the roof structure (full attic conversions can be included under F9 Other) • roof strengthening • lateral restraint, spreading of roof structure • the use and misuse of the roof space • wood rot and wood boring beetle • party walls – lack of fire breaks, gaps between dwellings • flooring in loft • possible asbestos containing materials • access into loft • ventilation in roof space • level of insulation • Birds, bats, vermin, insects, etc. **Flat roof structure** • access/restrictions to flat roof structure • sample moisture meter readings • ventilation to flat roof void • wood rot and wood boring insect (Services in the roof space to be included under appropriate elements in section G) **Links to other sections** • *cross reference to other affected elements* • *issues for legal adviser (section I)* • *risks (section J)*	

Property address/Postcode:

Section F – room-by-room supplement	
Room reference and location:	

Checklist
• ceiling
• internal walls and partitions
• floors
• fireplaces, chimney breasts, etc.
• woodwork:
   – *doors*
   – *windows*
   – *skirtings, etc.*
• sanitary fittings (baths, showers, basins, etc.)
• built-in fittings
 	
Services
• electrical
• water supply
• water heating
• heating
• drainage
	Comments

Room reference and location:	

Checklist
• ceiling
• internal walls and partitions
• floors
• fireplaces, chimney breasts, etc.
• woodwork:
   – *doors*
   – *windows*
   – *skirtings, etc.*
• sanitary fittings (baths, showers, basins, etc.)
• built-in fittings
 	
Services
• electrical
• water supply
• water heating
• heating
• drainage
	Comments

Property address/Postcode:

F9 Other	Description of construction and condition
Checklist This section is for anything internal that does not sit comfortably in earlier sections in section F. This could typically include: • **Roof space conversion** – include here where a total conversion has been carried out. To give a complete appraisal of the conversion, include all elements here. For example: – *structural aspects – alteration of roof, strengthening of ceilings and other floors* – *access to room – adequacy and safety of staircase, etc.* – *thermal insulation – including ventilation to roof spaces* – *fire escape route – fire resistance of partitions, ceilings, doors, etc.* – *regulation and permissions – building regulation, planning permission, listed building/conservation, etc.* • **Cellar/basements** – where the cellar/basements are used for habitable purposes, to give a complete appraisal, include all elements here. For example: – *structural aspects – removal of partitions, strengthening of ceilings and floors* – *access to room – adequacy and safety of staircase, etc.* – *fire escape and fire safety – additional fire resistance to ceiling, inner room issues, etc.* – *daylight and ventilation to rooms and spaces – compare with building regs* – *dampness – tanking issues, guarantees, etc.* • **Flats** – to include those internal common parts of a block: – *internal hallways, landings and stairs* – *doors and windows* – *built-in cupboards, meter cupboards, etc.* – *asbestos containing materials* – *asbestos management plan* **Links to other sections** • *cross reference to other affected elements* • *issues for legal adviser (section I)* • *risks (section J)*	

Property address/Postcode:

Section G - services	
Restrictions on inspection	
Include details of any restrictions on the inspection of the services:	
G1 Electricity	**Description of construction and condition**
Checklist • description and general condition • evidence of installation, alteration or maintenance certification (e.g. Part P approval, minor works cert., periodic inspections and reports, etc. **Main issues** (broad indicators only): • is there a mains supply • on-peak/off-peak supply • location of the meter and consumer unit/fuse board • separate from gas meter • is there a RCD/MCB • equipotential bonding in the usual places • type and condition of visible wiring • condition of a sample of the range of light fittings and switch gear • type and condition of fixed electrical appliances including heaters, storage radiators, electric showers, instant water heaters, etc. (see also G5 water heating) • bath and shower rooms – nature and suitability of electrical fittings • external installations – garages, outbuildings, external sockets, garden lighting, water feature pumps, etc. • external installations – covered by RCD **Links to other sections** • *cross reference to other affected elements* • *issues for legal adviser (section I)* • *risks (section J)*	**The following general advice will be printed in every report:** Safety warning: Periodic inspection and testing of electrical installations is important to protect your home from damage and to ensure the safety of the occupants. Guidance published by the Institution of Electrical Engineers recommends that inspections and testing are undertaken at least every ten years and on change of occupancy. All electrical installation work undertaken after 1 January 2005 should be identified by an Electrical Installation Certificate.

Property address/Postcode:

G2 Gas/Oil	Description of construction and condition

Please note: this section covers the gas and oil supply pipework from the utility company's supply (or the storage vessel) through to the heating appliance itself. All matters relating to the appliance are to be reported under G4 Heating.
 	
Checklist
• description and general condition
• evidence of installation, alteration or maintenance certification (e.g. Gas Safe Registered engineer reports, OFTEC certs., service agreement, etc.)
• **mains gas installations:**
• does it have a mains supply and is it connected
• meter location and condition – is it positioned on an escape route
• the route, nature and condition of the pipework – well fixed and supported
• have there been alterations and are they properly certified
• **LPG installations:**
• are the storage cylinders in a suitable location:
  – *boundaries*
  – *proximity to buildings (especially basements), other storage tanks*
• condition of tank, covered with foliage, etc
• the route, nature and condition of the pipework
• have there been alterations, and are they properly certified?
• **oil installations:**
• is this satisfactory, including:
• the position of tank:
  – *proximity to dwelling, boundaries, watercourses, drains, etc.*
  – *seen from filling point*
• nature of the tank:
  – *internal or external bunded*
  – *condition of tank*
 	
Links to other sections
• *cross reference to other affected elements*
• *issues for legal adviser (section I)*
• *risks (section J)*
	The following general advice will be printed in every report:
	Safety Warning: Regular inspection, testing, maintenance and servicing of all gas/oil appliances and equipment should be undertaken by a registered 'competent person' and in accordance with the manufacturer's instructions. This is important to ensure that such equipment is working correctly to minimise the risk of fire and carbon monoxide poisoning as well as leakages of carbon dioxide and other greenhouse gases to the atmosphere. For further advice contact Gas Safe Register for gas installations, and OFTEC for oil installations.

Other comments

Property address/Postcode:

G3 Water	Description of construction and condition

Checklist
General description of the system and condition including:
• the piped water supply from the boundary of the property:
  – *pipe material (lead?)*
  – *external stop valve*
  – *depth of pipe?*
  – *water meter*
  – *where does it enter building?*
• Internally, where the rising main appears:
  – *stop valve – accessibility*
  – *insulation (especially underfloor and in roof space)*
  – *material – lead? dissimilar materials?*
  – *support to pipework*
• cold water storage and expansion tanks:
  – *support to tank*
  – *tank type and condition*
  – *lid, thermal insulation*
  – *overflow*
• backflow prevention to:
  – *bidet, etc.*
  – *garden hose pipe,*
  – *irrigation system*
 	
Private water supplies
Main issues to include
• description and general condition
• private supply only or combined with mains
• nature of source
• has supply been checked and certified as wholesome source of supply
• shared supply with neighbours
 	
Links to other sections
• *cross reference to other affected elements*
• *issues for legal adviser (section I)*
• *risks (section J)*
	Please note: This element covers cold water supply service to combination boilers, water heaters, hot water cylinders but not the pipework beyond them (see G5). It covers cold water storage tanks and the down service beyond, but **not** the taps of appliances.

Other comments

Property address/Postcode:

G4 Heating	Description of construction and condition

Checklist
General description and condition of the heating system(s):
• evidence of certification for installation, maintenance, repair etc. (see also G2)
• boilers (including combination boilers that also provide hot water)
• gas and oil fires and heaters
• electric heating (but not the electricity supply to them
• heat pumps
• other forms (describe)
• heat distribution pipework
• heat emitters – radiators, panel heaters, under-floor heating pipework, etc.
• ancillary equipment including:
  – *heating controls*
  – *expansion vessels and safety controls*
• ventilation issues for heating appliances(where appropriate)
Please note: free standing solid fuel heating sources (incl. Aga) should be covered here. Where in fireplace include in F5 but distribution pipework, etc. covered under this element.
 	
Links to other sections
• *cross reference to other affected elements*
• *issues for legal adviser (section I)*
• *risks (section J)*
	Please note: Most heating appliances will be subject regulations governing installation of gas, oil, solid fuel and electric appliances previously described. Cross reference to appropriate section where certification is required.

G5 Water heating	Description of construction and condition
Checklist Include the hot water heating appliance where it provides hot water only, for example: • gas water heater • multi-point water heater • electric under-counter heater • instantaneous electric water heater; • solar water/PV heating Other parts of hot water system to include: • the hot water cylinder • insulation to tank and primary pipework • expansion pipe • immersion heaters • cylinder thermostats • all distribution pipework between cylinder and the appliances (but not the taps) • the distribution pipework between water heating source and the various appliances **Links to other sections** • *issues for legal adviser (section I)* • *risks (section J)*	**Please note:** Where hot water provided by combination heating, hot water boilers, heat pumps, etc. include under G4. Evidence of certification if not covered in G2/G4.

Property address/Postcode:

G6 Drainage	Description of construction and condition

Checklist
General description and condition of drainage system.
Above ground drainage:
• traps and wastes from sinks, washing machines, dishwashers, baths, bidets, WCs, showers, wash hand basin, etc.
• pipe material, condition, leaks
• length of wastes, cleaning access
• slope of branch connections
• soil and vent pipes, two pipe systems, etc.
• proximity of stack to windows
• stub stacks
• air admittance valves, etc.
Below ground drainage:
• gullies – type, condition, grid, blocked
• clearing access to drains
• inspection chambers:
  – *position*
  – *cover/condition*
  – *walls/benching/channel*
  – *standard of connections*
  – *Bldg regs for new work*
  – *interceptor*
  – *blocked*
• underground drains:
  – *type*
  – *building over drains*
  – *trees close by?*
  – *depth and approx fall?*
• design of system:
  – *separate soil and surface*
  – *combined system*
  – *inappropriate connections*
  – *join with other properties – private sewer?*
Private drainage including:
• design and type:
  – *cess pit*
  – *septic tank (two chamber)*
  – *small sewerage treatment system (e.g. Klargester, etc)*
• Other types:
  – *soakaway*
  – *reed beds*
  – *composting toilets.*
• Evidence of maintenance, permission to discharge, etc.
• Shared system
• Approx. size? Adequacy
• Evidence of misuse
  – *RWP connection*
  – *Extra WCs to small tank*
 	
Links to other sections
• *cross reference to other affected elements*
• *issues for legal adviser (section I)*
• *risks (section J)*	

Property address/Postcode:

G7 Common services	Description of construction and condition
Checklist General description and condition of the facilities to the block General comments in relation to: • refuse chutes, bin stores, etc. • lifts – passenger and goods • entry-phone systems • CCTV • common areas – heating and lighting • communal heat and power • renewable energy systems **Links to other sections** • *cross reference to other affected elements* • *issues for legal adviser (section I)* • *risks (section J)*	

Other comments

Property address/Postcode:

Section H – Grounds (including shared areas for flats)

Restrictions on inspection
Include details of any restrictions on the inspection of the services:

H1 Garage	Description of construction and condition

Checklist
General description, location, and condition including all elements and services, including:
• functionality:
  – *size for modern cars*
  – *access to entrance*
• condition:
  – *structure*
  – *walls/cladding*
  – *floor (inspection pits?)*
  – *roof – leaks, fragile covering*
  – *asbestos containing materials*
  – *windows/doors*
• services:
  – *electrics (RCD protection?)*
  – *heating*
  – *water*
  – *drainage*
• unauthorised use – commercial use, office, habitable, etc.
• conservation area, listed buildings issues
Please note: Integral garages should be including in the main report (mostly section F) under the elemental rating system.
 	
Links to other sections
• *issues for legal adviser (section I)*
• *risks (section J)*	

H2 Other	Description of construction and condition

Checklist
Other permanent outbuildings:
• summer houses
• permanent greenhouses that are to remain
• follies
• substantial garden sheds that are to remain
• leisure buildings (but not the leisure facilities inside)
• for each of the above:
  – *general description and condition including roof, walls, floors, windows, doors, etc.*
  – *services*
  – *electrics (RCD protection?)*
  – *heating*
  – *water*
  – *drainage*
• unauthorised use – commercial use, office, habitable, etc.
• conservation area, listed building issues
 	
Links to other sections
• *issues for legal adviser (section I)*
• *risks (section J)*	

Property address/Postcode:

H3 General	Description of construction and condition
Checklist To include a general description of the grounds and/or site including: • boundaries – condition and likely ownership, changed positions, etc. • retaining walls – construction, type, condition, etc. • terraces and patios • conservation and listed building issues • gardens – general description, approx. size, landscape/garden design features • drives and paths – type and condition • steps and changes of level – type, construction safety • trees – type, location to buildings and drains, maintenance, safety • ponds, lakes, reed beds, drainage fields, water courses, etc. – type, size, condition, ownership implications, safety • rights of way/easements across site/wayleaves, etc. • wider location/environment issues where impinge within boundaries, e.g. contamination, HV pylons, HV sub stations, landslip, etc. **Links to other sections** • *issues for legal adviser (section I)* • *risks (section J)*	

Other comments

Property address/Postcode:

Checklists for room-by-room supplement

Section F2 Ceilings

Checklist
- description of type and condition including cracking
- safety issues – stability of ceiling, bulging, etc.
- cornices/centrepieces other decorative features
- finishes, including decorations (heavy lining paper, textured coating, etc.)
- possible asbestos containing materials, textured decorative coatings, asbestos board, etc.
- Listed building features and issues

Links to other sections
- *cross reference to other affected elements*
- *issues for legal adviser (section I)*
- *risks (section J)*

Section F3 Internal walls and partitions

Checklist
- description of type and condition
- loading or non-load-bearing partitions
- structural movement, cracking, distortion, binding doors, etc.
- dampness to internal walls including condensation mould growth
- non-robust partitions, poor sound insulation, low level of fire resistance, etc.
- possible asbestos containing material, asbestos board, textured coating, etc.
- finishes including decorations
- wall tiling (including splash back wall tiling)
- dado panelling
- structural alterations – through lounge, new door openings, level of support, etc

Links to other sections
- *cross reference to other affected elements*
- *issues for legal adviser (section I)*
- *risks (section J)*

Section F4 Floors

Checklist
- description of type and condition (floor by floor)
- solid floor:
  - *cracking*
  - *settlement*
  - *swelling*
- suspended timber floor:
  - *stamp test and vibration*
  - *access hatch?*
  - *ventilation to sub floor void*
- description of floor finishes and their condition
- dampness to all floor types
- wood boring insects
- wood rot

Links to other sections
- *cross reference to other affected elements*
- *issues for legal adviser (section I)*
- *risks (section J)*

Section F5 Fireplaces, etc.

Checklist
Description of fireplaces, chimney breasts and flues and their condition
- **open, solid fuel fireplaces**:
  - *flue/chimney maintenance sweeping*
  - *lining*
  - *Part J approvals*
  - *noncombust. hearth*
  - *certificates, invoices, etc.*
  - *fixed external ventilation*
  - *hearth*
- **chimney breasts**:
  - *open or blocked off*
  - *if blocked – ventilation to flue*
- terminal at top of flue at chimney level
- condition of chimney breast in roof space:
  - *structural alterations unsupported chimney breast*
  - *dampness to base and face of chimney breasts*
- **independent open vertical flues**:
  - *description and fixing type, and condition*
  - *suitability of terminal position*
- proximity of flue to combustible materials, etc.
- possible asbestos materials
- listed building issues

Links to other sections
- *cross reference to other affected elements*
- *issues for legal adviser (section I)*
- *risks (section J)*

F6 Built-in fittings

Checklist
Description, adequacy and general condition

- **kitchen units, cupboards and worktops:**
  - *the waterproof sealant but not the splash back tiling (see F3)*
- **kitchen sinks:**
  - *the taps (but not the water pipes to them – G3 Water)*
  - *the slotted wastes (but not the trap or the waste pipe: G6 Drainage)*
- dampness affecting the features described above
- wood-boring insects affecting the features described above
- operation and functionality – operable drawers, damaged hinges, etc.
- age and modernity – appeal to purchasers

Links to other sections
- *cross reference to other affected elements*
- *issues for legal adviser (section I)*
- *risks (section J)*

Property address/Postcode:

F7 Woodwork

Checklist
Description and general condition:
- internal doors, frames/linings, architraves, etc.
- internal parts of windows – jamb linings, sill boards, architraves, shutters, etc.
- built-in cupboards – for example within chimney alcoves, meter cupboards, understairs
- skirtings, dado rails, picture rails
- staircases – treads and risers, strings, balusters, handrails, newel posts:
  - *safety hazards – missing balusters, broken treads, etc.*
- finishes to features above including decorations
- dampness affecting the features described above
- wood rot and wood-boring beetle affecting the features described above

Links to other sections
- *cross reference to other affected elements*
- *issues for legal adviser (section I)*
- *risks (section J)*

F8 Bathroom fittings
Checklist
Description, adequacy and general condition:
- baths, bidets, wash hand basins, their taps and slotted wastes but **not** the water pipes to them (G3 and G5), and **not** the traps and waste pipe (G6)
- shower trays, shower cubicles but **not** the mixer valve (G3) or electric shower (G1)
- WCs, their cisterns and overflows. To include junction between the WC outgo and the drainage pipe but not the drainage pipe itself (G6)
- bath panels and other built-in fittings associated with the sanitary appliances
- the sealant between the appliance and the adjacent surfaces but not the splash back tiles (see F3)
- leaks and dampness caused by the bathroom fittings
- wood rot and wood-boring insects affecting the features described above

Links to other sections
- *cross reference to other affected elements*
- *issues for legal adviser (section I)*
- *risks (section J)*

F9 Other

Checklist
This section is for anything internal that does not sit comfortably earlier sections in section F. This could typically include:

- **Roof space conversion** – include here where a total conversion has been carried out. To give a complete appraisal of the conversion, include all elements here. For example:
  - *structural aspects – alteration of roof, strengthening of ceilings and other floors*
  - *access to room – adequacy and safety of staircase, etc.*
  - *thermal insulation – including ventilation to roof spaces*
  - *fire escape route – fire resistance of partitions, ceilings, doors, etc.*
  - *regulation and permissions – building regulation, planning permission, listed building/conservation, etc.*

- **Cellar/basements** – where the cellar/basements are used for habitable purposes, to give a complete appraisal, include all elements here. For example:
  - *structural aspects – removal of partitions, strengthening of ceilings and floors*
  - *access to room – adequacy and safety of staircase, etc.*
  - *fire escape and fire safety – additional fire resistance to ceiling, inner room issues, etc.*
  - *daylight and ventilation to rooms and spaces – compare with building regs*
  - *dampness – tanking issues, guarantees, etc.*

- **Flats** – to include those internal common parts of a block:
  - *internal hallways, landings and stairs*
  - *doors and windows*
  - *built-in cupboards, meter cupboards, etc.*
  - *asbestos containing materials*
  - *asbestos management plan*

Links to other sections
- *cross reference to other affected elements*
- *issues for legal adviser (section I)*
- *risks (section J)*

APPENDIX 1 *(c) HBR site notes (for dictation)*

RICS HomeBuyer Report – Site notes (for dictation)

Page 1 of 13

Surveyor's name		Company/address	
Date of inspection		Report reference	
Appointment arrangements and time of arrival		Time of departure	
Full address of the property Postcode:		Weather conditions	

Circumstances of inspection

Owner occupied / Tenanted / Vacant Comments:	Furnished / part furnished / empty / derelict Comments:
Access to HIP YES / NO Comments:	**EPC details** Name of DEA: Reference of certificate: Energy efficiency rating: Environmental impact rating:

Access to Property Information Questionnaire   YES / NO
Comments on content: (e.g. maintenance/safety checks of services, type of documentation checked; declared defects and alterations of property, etc.)

Type of property (please indicate)

House	Bungalow	Chalet	Flat	Maisonette	Shop/ commercial
Detached	Semi-detached	Terraced	End-terraced	Enclosed terr.	Other

Construction (broad indication of roof, walls, floor, etc.)

Date of construction/extension/conversion (describe evidence)

A Surveyor's Guide to the RICS HomeBuyer Service

Property address/Postcode: Page 2 of 14

Accommodation

Floor	Living rooms	Bedrooms	Bath/ shower	Separate toilet	Kitchen	Utility room	Conservatory	Other (name / details)
Lower ground								
Ground								
First								
Second								
Third								
Other								
Roof space								

Main services:	Drainage	Gas	Electricity	Water
General comments on services and other energy sources (wind turbine, PV, etc.):				

Flats and maisonettes:

Which floor?		How many floors in block?	
No. of flats in block?		Converted?	
Purpose built?		Lift (describe)	
Commercial use? (describe, type, location, and % of block)			

Central heating:	Gas (mains)	Gas (LPG)	Electricity	Solid fuel
	Oil	Other (describe)		

Health and safety hazards, e.g.	**Commentary** (This section should give a short account of possible health and safety hazards in the property at the time of inspection. This should include an assessment of the level of risk (low; medium; high) and what action was taken to reduce the risk.)
• parking/road safety • empty property • access • animals • occupants • asbestos fibres • vermin, rats, pigeons, etc. • voids/holes/unsafe areas • tripping hazards • ladders and fixed access equipment • services – electrics, exposed wires, gas, LPG, plumbing, drains, etc. • weather – high winds, etc. • any other issues	

Property address/Postcode: Page 3 of 14

Checklists for dictation –
Outside the property and general details about location

Outside the property and the location

Checklist
- location of property
- neighbourhood
- adjoining uses
- local nuisances
- local planning applications
- parking/permits/access arrangements
- roads/rail/airports

Links to other sections
- *issues for legal adviser (section I)*
- *risks (section J)*

Facilities

Checklist
- shops
- schools
- transport – bus routes, railway station, etc.
- local services – doctors, etc.

Links to other sections
- *issues for legal adviser (section I)*
- *risks (section J)*

Local environment

Checklist
- flood risks
- radon
- EMFs
- mining activity
- contaminated/brownfield sites
- clay sub soils
- coastal erosion
- landslip area
- mundic area
- conservation area

Links to other sections
- *issues for legal adviser (section I)*
- *nsks (section J)*

Other comments

Property address/Postcode:

Site plan/External ....

Checklist

North point
Main outbuildings
Boundaries
Trees
Drain lines
Inspect chambers
Retaining walls

Property address/Postcode:

Floor plan ....

Checklist

North point

Dampness readings

Restrictions (furniture, floor cover, etc.)

Dimensions

Room heights

Property address/Postcode:

Floor plan ....

Checklist
North point
Dampness readings
Restrictions (furniture, floor cover, etc.)
Dimensions
Room heights

Property address/Postcode:

Other comments and observations

Property address/Postcode:

Checklists for dictation – Section E External

Section E1
Chimney stacks

Checklist
- description, number and location
- flue terminals
- chimney pots
- flaunching
- condition of stacks (leaning, cracked, sulfate attack, etc.)
- pointing, render, other finishes
- aerials and satellite dishes
- flashings and soakers at the junction with the roof covering (but not including the roof covering)
- any party wall issues arising from chimney condition
- listed building/conservation areas issues

Links to other sections
- cross reference to other affected elements
- issues for legal adviser (section I)
- risks (section J)

Section E2
Roof coverings

Pitched roof
- type of covering and general condition including:
  – the slope/pitch of the roof
  – fixing of tiles, slates
  – temp. repairs
- roofing felt/sarking boarding – presence, type, condition
- ridges and hips tiles
- verge and eaves details
- open valleys
- valley gutters ('butterfly' roofs), parapets gutters, their linings and outlets
- lead flashings to up-stands, dormers, etc. (but not the flashings to the chimney)
- roof lights/roof windows
- dormers (usually including the flat or pitched over the feature)
- any party wall issues arising from the roof condition
- listed building/conservation area issues

Flat roof
- type of covering and condition including:
  – the falls on the roof
  – pooling of water
  – temp. repairs
  – solar protection
- lead flashings to up-stands (but not any chimney flashings)
- other penetrations (e.g. SVP, vertical flues, etc.)
- verge and eaves details
- any integral gutters
- roof lights and upstands
- any party wall issues arising from the roof condition;
- listed building/conservation areas

Links to other sections
- cross reference to other affected elements
- issues for legal adviser (section I)
- risks (section J)

Section E3
RW gutters and downpipes

Checklist
- type of gutters, gutter supports and condition including:
  – falls to gutter
  – position and adequacy of downpipes, hopper, etc.
  – debris/blockage
  – leaks and damage to other parts of building
- swan necks and off-sets
- rainwater downpipes
- hopper heads
- rainwater shoes, etc.
- listed building/conservation areas;
- any party issues with gutters and downpipes, e.g. shared roof drainage, etc.

Links to other sections
- cross reference to other affected elements
- issues for legal adviser (section I)
- risks (section J)

Section E4 Main walls

Checklist
- type, thickness and general condition of the wall(s) including:
  – wall ties or sulfate attack
  – wall tie replacement
  – pointing
  – rendering/external finishes
  – cracking/distortion
  – bulging/verticality
  – efflorescence
- cladding:
  – vertical tiles;
  – PVC
  – timber
- damp proof course – type, position, and condition
- evidence if injection holes
- relationship with external ground levels
- presence of dampness to the inside face of the outside wall from rising and penetrating dampness and condensation
- the type and condition of the lintels, jambs and sills
- parapet walls and copings including party up-stand walls
- timber frame walls
- type and extent of insulation to the walls
- any party wall issues arising from the wall condition
- any listed building/conservation areas issues

Links to other sections
- cross reference to other affected elements
- issues for legal adviser (section I)
- risks (section J)

Section E5 Windows

Checklist
- the type and general condition of the windows
- orientation of windows
- opening of window sample
- single/double/secondary glazing
- seals to frame
- damp penetration around frames
- draughts/trickle vents
- seals to double glazed units
- replacement window issues
- escape in the case of fire
- low level glass/safety glass
- security
- listed building/conservation area issues

Links to other sections
- cross reference to other affected elements
- issues for legal adviser (section I)
- risks (section J)

Property address/Postcode:

Section E6
Outside doors
(incl. patio doors)

Checklist
- type/condition of door and frame
- level threshold? If level access, exclusion of water, etc.
- glazing and safety glass issues, safety glass, etc.
- security
- draughts/damp penetration around frame
- seals to double glazed units
- listed building/conservation area issues

Links to other sections
- *cross reference to other affected elements*
- *issues for legal adviser (section I)*
- *risks (section J)*

Section E7 Conservatories and porches

Checklist
- type and condition
- inappropriate use as habitable room
- single/double glazing
- safety - structural stability and safety glazing
- snow/debris guards to edge of dwelling roof
- separation from rooms of the main dwelling
- waterproofing at junction with dwelling
- opening of sample of windows
- security of windows/doors
- effect on escape in the case of fire from the upper windows of the dwelling
- guarantee/warranty
- building regulation and/or planning approval
- listed building/conservation area issues

Links to other sections
- *cross reference to other affected elements*
- *issues for legal adviser (section I)*
- *risks (section J)*

Section E8
Other joinery and finishes

Checklist
- type and condition of joinery including:
  - *signs of rot*
  - *state of the decorations*
- fascias, soffits and eaves details generally
- bargeboards and verge details;
- ornamental joinery (finials, mock Tudor panelling, etc.);
- timber and PVC cladding fixed to main walling
- possible asbestos containing materials

Links to other sections
- *cross reference to other affected elements*
- *issues for legal adviser (section I)*
- *risks (section J)*

Section E9 Other

Checklist
- roof terraces – useable flat roof areas over other accommodation below
- balconies
- large bay windows where distinctive and different in nature to the construction of the main walls
- external stairways and fire escapes, etc.

For each feature, include all aspects of the feature, e.g.:
- structural aspects – fixings, support, stability, etc.
- the walls, floor and roof (if present) to the feature
- railings, handrails, etc.

Links to other sections
- *cross reference to other affected elements*
- *issues for legal adviser (section I)*
- *risks (section J)*

Other issues and comments
Include any other external matters here.

Property address/Postcode:

Checklists for dictation – Section F Internal

Section F1
Roof space and structure

Checklist
Pitched roofs
- description of design, construction and condition of the roof structure
- sample moisture readings
- alteration of the roof structure (full attic conversions can be included under F9 Other)
- roof strengthening
- lateral restraint, spreading of roof structure
- the use and misuse of the roof space
- wood rot and wood boring beetle
- party walls – lack of fire breaks, gaps between dwellings
- flooring in loft
- possible asbestos containing materials
- access into loft
- ventilation in roof space
- level of insulation
- Birds, bats, vermin, insects, etc.

Flat roof structure
- access/restrictions to flat roof structure
- sample moisture meter readings
- ventilation to flat roof void
- wood rot and wood boring insect

(Services in the roof space to be included under appropriate elements in section G)

Links to other sections
- *cross reference to other affected elements*
- *issues for legal adviser (section I)*
- *risks (section J)*

Section F2 Ceilings

Checklist
- description of type and condition including cracking
- safety issues – stability of ceiling, bulging, etc.
- cornices/centrepieces other decorative features
- finishes, including decorations (heavy lining paper, textured coating, etc.)
- possible asbestos containing materials, textured decorative coatings, asbestos board, etc.
- Listed building features and issues

Links to other sections
- *cross reference to other affected elements*
- *issues for legal adviser (section I)*
- *risks (section J)*

Section F3 Internal walls and partitions

Checklist
- description of type and condition
- loading or non-load-bearing partitions
- structural movement, cracking, distortion, binding doors, etc.
- dampness to internal walls including condensation mould growth
- non-robust partitions, poor sound insulation, low level of fire resistance, etc.
- possible asbestos containing material, asbestos board, textured coating, etc.
- finishes including decorations
- wall tiling (including splash back wall tiling)
- dado panelling
- structural alterations – through lounge, new door openings, level of support, etc

Links to other sections
- *cross reference to other affected elements*
- *issues for legal adviser (section I)*
- *risks (section J)*

Section F4 Floors

Checklist
- description of type and condition (floor by floor)
- solid floor:
  – *cracking*
  – *settlement*
  – *swelling*
- suspended timber floor:
  – *stamp test and vibration*
  – *access hatch?*
  – *ventilation to sub floor void*
- description of floor finishes and their condition
- dampness to all floor types
- wood boring insects
- wood rot

Links to other sections
- *cross reference to other affected elements*
- *issues for legal adviser (section I)*
- *risks (section J)*

Section F5 Fireplaces, etc.

Checklist
Description of fireplaces, chimney breasts and flues and their condition
- **open, solid fuel fireplaces**:
  – *flue/chimney maintenance sweeping*
  – *lining*
  – *Part J approvals*
  – *noncombust. hearth*
  – *certificates, invoices, etc.*
  – *fixed external ventilation*
  – *hearth*
- **chimney breasts:**
  – *open or blocked off*
  – *if blocked – ventilation to flue*
- terminal at top of flue at chimney level
- condition of chimney breast in roof space:
  – *structural alterations unsupported chimney breast*
  – *dampness to base and face of chimney breasts*
- **independent open vertical flues:**
  – *description and fixing type, and condition*
  – *suitability of terminal position*
- proximity of flue to combustible materials, etc.
- possible asbestos materials
- listed building issues

Links to other sections
- *cross reference to other affected elements*
- *issues for legal adviser (section I)*
- *risks (section J)*

Property address/Postcode:

Section F6 Built-in fittings

Checklist
Description, adequacy and general condition

- **kitchen units, cupboards and worktops:**
  - *the waterproof sealant but not the splash back tiling (see F3)*
- **kitchen sinks:**
  - *the taps (but not the water pipes to them – G3 Water)*
  - *the slotted wastes (but not the trap or the waste pipe: G6 Drainage)*
- dampness affecting the features described above
- wood-boring insects affecting the features described above
- operation and functionality – operable drawers, damaged hinges, etc.
- age and modernity – appeal to purchasers

Links to other sections
- *cross reference to other affected elements*
- *issues for legal adviser (section I)*
- *risks (section J)*

Section F7 Woodwork

Checklist
Description and general condition:
- internal doors, frames/linings, architraves, etc.
- internal parts of windows – jamb linings, sill boards, architraves, shutters, etc.
- built in cupboards – for example within chimney alcoves, meter cupboards, understairs
- skirtings, dado rails, picture rails
- staircases – treads and risers, strings, balusters, handrails, newel posts:
  - *safety hazards – missing balusters, broken treads, etc.*
- finishes to features above including decorations
- dampness affecting the features described above
- wood rot and wood-boring beetle affecting the features described above

Links to other sections
- *cross reference to other affected elements*
- *issues for legal adviser (section I)*
- *risks (section J)*

Section F8 Bathroom fittings

Checklist
Description, adequacy and general condition:
- baths, bidets, wash hand basins, their taps and slotted wastes but **not** the water pipes to them (G3 and G5), and **not** the traps and waste pipe (G6)
- shower trays, shower cubicles but **not** the mixer valve (G3) or electric shower (G1)
- WCs, their cisterns and overflows. To include junction between the WC outgo and the drainage pipe but not the drainage pipe itself (G6)
- bath panels and other built-in fittings associated the sanitary appliances
- the sealant between the appliance and the adjacent surfaces but not the splash back tiles (see F3)
- leaks and dampness caused by the bathroom fittings
- wood rot and wood-boring insects affecting the features described above

Links to other sections
- *cross reference to other affected elements*
- *issues for legal adviser (section I)*
- *risks (section J)*

Section F9 Other

Checklist
This section is for anything internal that does not sit comfortably in earlier sections in section F. This could typically include:

- **Roof space conversion** – include here where a total conversion has been carried out. To give a complete appraisal of the conversion, include all elements here. For example:
  - *structural aspects – alteration of roof, strengthening of ceilings and other floors*
  - *access to room – adequacy and safety of staircase, etc.*
  - *thermal insulation – including ventilation to roof spaces*
  - *fire escape route – fire resistance of partitions, ceilings, doors, etc.*
  - *regulation and permissions – building regulation, planning permission, listed building/conservation, etc.*

- **Cellar/basements** – where the cellar/basements are used for habitable purposes, to give a complete appraisal, include all elements here. For example:
  - *structural aspects – removal of partitions, strengthening of ceilings and floors*
  - *access to room – adequacy and safety of staircase, etc.*
  - *fire escape and fire safety – additional fire resistance to ceiling, inner room issues, etc.*
  - *daylight and ventilation to rooms and spaces – compare with building regs*
  - *dampness – tanking issues, guarantees, etc.*

- **Flats** – to include those internal common parts of a block:
  - *internal hallways, landings and stairs*
  - *doors and windows*
  - *built-in cupboards, meter cupboards, etc.*
  - *asbestos containing materials*
  - *asbestos management plan*

Links to other sections
- *cross reference to other affected elements*
- *issues for legal adviser (section I)*
- *risks (section J)*

Property address/Postcode:

Checklists for dictation – Section G Services

Section G1 Electricity

Checklist
- description and general condition
- evidence of installation, alteration or maintenance certification (e.g. Part P approval, minor works cert., periodic inspections and reports, etc.

Main issues (broad indicators only):
- is there a mains supply
- on-peak/off-peak supply
- location of the meter and consumer unit/fuse board
- separate from gas meter
- is there a RCD/MCB
- equipotential bonding in the usual places
- type and condition of visible wiring
- condition of a sample of the range of light fittings and switch gear
- type and condition of fixed electrical appliances including heaters, storage radiators, electric showers, instant water heaters, etc. (see also G5 water heating)
- bath and shower rooms – nature and suitability of electrical fittings
- external installations – garages, outbuildings, external sockets, garden lighting, water feature pumps, etc.
- external installations – covered by RCD

Links to other sections
- cross reference to other affected elements
- issues for legal adviser (section I)
- risks (section J)

Section G2 Gas/Oil

Please note: this section covers the gas and oil supply pipework from the utility company's supply (or the storage vessel) through to the heating appliance itself. All matters relating to the appliance are to be reported under G4 Heating.

Checklist
- description and general condition
- evidence of installation, alteration or maintenance certification (e.g. Gas Safe Registered engineer reports, OFTEC certs., service agreement, etc.)

- **mains gas installations:**
- does it have a mains supply and is it connected
- meter location and condition – is it positioned on an escape route
- the route, nature and condition of the pipework – well fixed and supported
- have there been alterations and are they properly certified
- **LPG installations:**
- are the storage cylinders in a suitable location:
  – boundaries
  – proximity to buildings (especially basements), other storage tanks
- condition of tank, covered with foliage, etc
- the route, nature and condition of the pipework
- have there been alterations, and are they properly certified?
- **oil installations:**
- is this satisfactory, including:
- the position of tank:
  – proximity to dwelling, boundaries, watercourses, drains, etc.
  – seen from filling point
- nature of the tank:
  – internal or external bunded
  – condition of tank

Links to other sections
- cross reference to other affected elements
- issues for legal adviser (section I)
- risks (section J)

Section G3 Water

Checklist
General description of the system and condition including:
- the piped water supply from the boundary of the property:
  – pipe material (lead?)
  – external stop valve
  – depth of pipe?
  – water meter
  – where does it enter building?
- Internally, where the rising main appears:
  – stop valve – accessibility
  – insulation (especially underfloor and in roof space)
  – material – lead? dissimilar materials?

– support to pipework
- cold water storage and expansion tanks:
  – support to tank
  – tank type and condition
  – lid, thermal insulation
  – overflow
- backflow prevention to:
  – bidet, etc.
  – garden hose pipe,
  – irrigation system

Private water supplies
Main issues to include
- description and general condition
- private supply only or combined with mains
- nature of source
- has supply been checked and certified as wholesome source of supply
- shared supply with neighbours

Links to other sections
- cross reference to other affected elements
- issues for legal adviser (section I)
- risks (section J)

Section G4 Heating

Checklist
General description and condition of the heating system(s):
- evidence of certification for installation, maintenance, repair, etc. (see also G2)
- boilers (including combination boilers that also provide hot water)
- gas and oil fires and heaters
- electric heating (but not the electricity supply to them
- heat pumps
- other forms (describe)
- heat distribution pipework
- heat emitters – radiators, panel heaters, under-floor heating pipework, etc.
- ancillary equipment including:
  – heating controls
  – expansion vessels and safety controls
- ventilation issues for heating appliances(where appropriate)
Please note: free standing solid fuel

Property address/Postcode:

heating sources (incl. Aga) should be covered here. Where in fireplace include in F5 but distribution pipework, etc. covered under this element.

Links to other sections
- *cross reference to other affected elements*
- *issues for legal adviser (section I)*
- *risks (section J)*

Section G5 Water heating

Checklist
Include the hot water heating appliance where it provides hot water only, for example:
- gas water heater
- multi-point water heater
- electric under-counter heater
- instantaneous electric water heater;
- solar water/PV heating

Other parts of hot water system to include:
- the hot water cylinder
- insulation to tank and primary pipework
- expansion pipe
- immersion heaters
- cylinder thermostats
- all distribution pipework between cylinder and the appliances (but not the taps)
- the distribution pipework between water heating source and the various appliances

Links to other sections
- *issues for legal adviser (section I)*
- *risks (section J)*

Section G6 Drainage

Checklist
General description and condition of drainage system.

Above ground drainage:
- traps and wastes from sinks, washing machines, dishwashers, baths, bidets, WCs, showers, wash hand basin, etc.
- pipe material, condition, leaks
- length of wastes, cleaning access
- slope of branch connections
- soil and vent pipes, two pipe systems, etc.
- proximity of stack to windows
- stub stacks
- air admittance valves, etc.

Below ground drainage:
- gullies – type, condition, grid, blocked
- clearing access to drains
- inspection chambers:
  – *position*
  – *cover/condition*
  – *walls/benching/channel*
  – *standard of connections*
  – *Bldg regs for new work*
  – *interceptor*
  – *blocked*
- underground drains:
  – *type*
  – *building over drains*
  – *trees close by?*
  – *depth and approx fall?*
- design of system:
  – *separate soil and surface*
  – *combined system*
  – *inappropriate connections*
  – *join with other properties – private sewer?*

Private drainage including:
- design and type:
  – *cess pit*
  – *septic tank (two chamber)*
  – *small sewerage treatment system (e.g. Klargester, etc)*
- Other types:
  – *soakaway*
  – *reed beds*
  – *composting toilets.*
- Evidence of maintenance, permission to discharge, etc.
- Shared system
- Approx. size? Adequacy
- Evidence of misuse
  – *RWP connection*
  – *Extra WCs to small tank*

Links to other sections
- *cross reference to other affected elements*
- *issues for legal adviser (section I)*
- *risks (section J)*

Section G7 Common services

Checklist
General description and condition of the facilities to the block

General comments in relation to:
- refuse chutes, bin stores, etc.
- lifts – passenger and goods
- enter-phone systems
- CCTV
- common areas – heating and lighting
- communal heat and power
- renewable energy systems

Links to other sections
- *cross reference to other affected elements*
- *issues for legal adviser (section I)*
- *risks (section J)*

A Surveyor's Guide to the RICS HomeBuyer Service

Property address/Postcode:

Checklists for dictation – Section H Grounds (incl. shared areas for flats)

Section H1 Garages

Checklist
General description, location, and condition including all elements and services, including:
- functionality:
  - *size for modern cars*
  - *access to entrance*
- condition:
  - *structure*
  - *walls/cladding*
  - *floor (inspection pits?)*
  - *roof – leaks, fragile covering*
  - *asbestos containing materials*
  - *windows/doors*
- services:
  - *electrics (RCD protection?)*
  - *heating*
  - *water*
  - *drainage*
- unauthorised use – commercial use, office, habitable, etc.
- conservation area, listed buildings issues

Please note: Integral garages should be including in the main report (mostly section F) under the elemental rating system.

Links to other sections
- *issues for legal adviser (section I)*
- *risks (section J)*

Section H2 Other outbuildings

Checklist
Other permanent outbuildings:
- summer houses
- permanent greenhouses that are to remain
- follies
- substantial garden sheds that are to remain
- leisure buildings (but not the leisure facilities inside)
- for each of the above:
  - *general description and condition including roof, walls, floors, windows, doors, etc.*
  - *services*
  - *electrics (RCD protection?)*
  - *heating*
  - *water*
  - *drainage*
- unauthorised use – commercial use, office, habitable, etc.
- conservation area, listed buildings issues

Links to other sections
- *issues for legal adviser (section I)*
- *risks (section J)*

Section H3 General

Checklist
To include a general description of the grounds and/or site including:
- boundaries – condition and likely ownership, changed positions, etc.
- retaining walls – construction, type, condition, etc.
- terraces and patios
- conservation and listed building issues
- gardens – general description, approx. size, landscape/garden design features
- drives and paths – type and condition
- steps and changes of level – type, construction safety
- trees – type, location to buildings and drains, maintenance, safety
- ponds, lakes, reed beds, drainage fields, water courses, etc. – type, size, condition, ownership implications, safety
- rights of way/easements across site/wayleaves, etc.
- wider location/environment issues where impinge within boundaries e.g. contamination, HV pylons, HV sub stations, landslip, etc.

Links to other sections
- *issues for legal adviser (section I)*
- *risks (section J)*

APPENDIX 2 *Sellers' questionnaire*

Sellers' questionnaire

Property address/Postcode:

About the property	Response
What date was the house built?	
Is the property constructed using any particular system or using non-traditional materials (e.g. concrete or steel framed, etc.)?	
If it is relatively new, do you know the name of the builder who constructed the house?	
What is the tenure – freehold or leasehold? If leasehold: <ul><li>How many years until expiry?</li><li>What is the ground rent?</li><li>What is the service charge?</li></ul>	
How long have you owned/leased the property?	
Is the property liable to flooding? If yes, when was the last flood?	
What fixtures or fittings are going to be removed?	
What chattels/possessions are going to remain?	

Legal issues	
Are there any rights of way over the property?	
Is there any shared access to the gardens or grounds?	
Do you know which boundaries you are responsible for maintaining?	
Is your property currently affected by neighbour or boundary disputes? If yes, please give brief details:	
Are there any outstanding party wall issues?	
Has the road to the property been adopted by the local authority or is it privately owned?	
Is the property a 'listed building' or is it in a conservation area?	

Property address/Postcode:

Are there any tree preservation orders on trees within the grounds?	
Are there any flying freeholds? (i.e. do any parts of your property protrude over your neighbour's)? If yes, please give the details.	
Do you know if there are any bats in the property?	
Does the property have any concealed access hatches or doorways that can be opened?	
Are there any parts of the property that cannot be inspected because of restricted access e.g. stored possessions, occupier affected by an illness, etc?	
Repairs and defects	
Have you carried out any structural alteration and/or repair work or added an extension, converted a loft, etc? If yes: • what did this involve? • what date did you do this? • did you obtain building regulation approval • and planning permission? • if yes to above, do you have the appropriate documentation?	
Do you have any guarantees/warranties for any of the repairs carried out?	
Is the property affected by any known current defects or other problems? If yes, please give full details.	
Have you improved the energy efficiency of the property? If yes, please give details	

Property address/Postcode: **Page 3 of 3**

Services	
What type of service systems does the property have: • electricity • mains water or a private water supply/mains drainage or a private sewerage system • mains gas, LPG, oil, solid fuel. Please give details	
When was the property last rewired?	
How old is the main boiler/source of heating?	
Do you have maintenance records for the service systems?	
If you have a cess pit or septic tank: • how often is it emptied? • where is the tank? • how big is it?	

Other notes

I have checked the above information and can confirm that it is accurate:

Signed:	Date:

APPENDIX 3 *Market valuation analysis*

Market valuation analysis

Page 1 of 1

Property address/Postcode:

Address:									Asking price: £			
									Sale Price: £			
Agent:									Date of sale:			
Type	Floors	Living	Cloaks	Beds	Baths	Gar.	C.htg	D.G.	Year	Size	Cond.	Location	Distance
										Smaller	Worse	Worse	
Extras										Similar	Similar	Similar	
										Larger	Better	Better	

Address:									Asking price: £			
									Sale Price: £			
Agent:									Date of sale:			
Type	Floors	Living	Cloaks	Beds	Baths	Gar.	C.htg	D.G.	Year	Size	Cond.	Location	Distance
										Smaller	Worse	Worse	
Extras										Similar	Similar	Similar	
										Larger	Better	Better	

Address:									Asking price: £			
									Sale Price: £			
Agent:									Date of sale:			
Type	Floors	Living	Cloaks	Beds	Baths	Gar.	C.htg	D.G.	Year	Size	Cond.	Location	Distance
										Smaller	Worse	Worse	
Extras										Similar	Similar	Similar	
										Larger	Better	Better	

Market valuation justification – matters affecting value
(please add comments)

Date:

APPENDIX 4 *HBR site notes (sample)*

RICS HomeBuyer Report – Site notes

Surveyor's name	*Phil Parnham*	**Company/address**	*Bluebox Partners* *Long Henry Row* *Hype Park, Coventry CV1*
Date of inspection	*25th December 2008*	**Report reference**	*00001*
Appointment arrangements and time of arrival	*Collect keys from agent at 10.00am 25/12/08 Arrived site 10.20.*	**Time of departure**	*Left property 12.10*
Full address of the property *52 Recession Avenue* *Coventry* Postcode: *CV1 1GY*		**Weather conditions** *Dry and sunny with cloudy intervals last two weeks – dry weather*	

Circumstances of inspection

Owner occupied / Tenanted / (Vacant)
Comments:

Furnished / part furnished / empty / derelict
Comments: *Still partly furnished with carpets in every room*

Access to HIP YES / NO
Comments: *No – the agent said that a HIP has not been assembled yet.*

EPC details
Name of DEA: *None available*

Reference of certificate: *Not done yet*

Energy efficiency rating:

Environmental impact rating:

Access to Property Information Questionnaire YES / NO
Comments on content: (e.g. maintenance/safety checks of services, type of documentation checked; declared defects and alterations of property, etc.)

Not applicable as no HIP available

Type of property (please indicate)

(House)	Bungalow	Chalet	Flat	Maisonette	Shop/commercial
Detached	Semi-detached	Terraced	(End-terraced)	Enclosed terr.	Other

Construction (broad indication of roof, walls, floor, etc.)

Solid brick walls with rough cast render to all elevations. Fair faced brick below dpc, mock cavity (fair faced) below ground floor bay window. Timber ground floors apart from kitchen + utility. Pitched + hipped roof, gable over bay, covered with tiles (last 10 years)

Date of construction/extension/conversion (describe evidence)

Based on knowledge of area 1930-35. Rear utility room not original but added – say 39-40 years ago.

Property address/Postcode: 52 Recession Ave, Coventry

Accommodation

Floor	Living rooms	Bedrooms	Bath/ shower	Separate toilet	Kitchen	Utility room	Conservatory	Other (name / details)
Lower ground								
Ground	2				1	1		
First		3	1					
Second								
Third								
Other								
Roof space								

Main services:	(Drainage)	(Gas)	(Electricity)	(Water)

General comments on services and other energy sources (wind turbine, PV, etc.):

None

Flats and maisonettes:

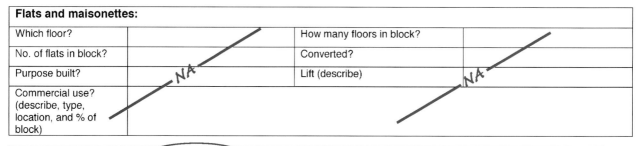

Which floor?		How many floors in block?	
No. of flats in block?		Converted?	
Purpose built?	NA	Lift (describe)	NA
Commercial use? (describe, type, location, and % of block)			

Central heating:	(Gas (mains))	Gas (LPG)	Electricity	Solid fuel
	Oil	Other (describe)		

Health and safety hazards, e.g. • parking/road safety • empty property • access • animals • occupants • asbestos fibres • vermin, rats, pigeons, etc. • voids/holes/unsafe areas • tripping hazards • ladders and fixed access equipment • services – electrics, exposed wires, gas, LPG, plumbing, drains, etc. • weather – high winds, etc. • any other issues	**Commentary** (This section should give a short account of possible health and safety hazards in the property at the time of inspection. This should include an assessment of the level of risk (low; medium; high) and what action was taken to reduce the risk.) – Empty property (med. risk) – told agent when keys returned. Told my office of whereabouts + expected time completion – Kept doors locked all times; checked for illegal occupants – Garden overgrown – tripping hazards, stout shoes – Loose boarding in loft; stepped on rear of visible joists only – Services all switched off – Asbestos content to garage roof – doors locked, did not enter.

Appendices

Property address/Postcode: 52 Recession Ave, Coventry | Page 3 of 26

Outside the property and the location

Checklist
- location of property
- neighbourhood
- adjoining uses
- local nuisances
- local planning applications
- parking/permits/access arrangements
- roads/rail/airports

Please see section H for detailed comments on garages, outbuildings, grounds.

Typical residential location, other similar properties mainly owner occ. But evidence of some renting. Plenty of on street parking, not much traffic during inspection. Gardens to front + rear, passage down side (shared) with locked gate across – who has key? Garage at rear of rear garden, door locked – can't get into rear access road. Is it usable?

Legal adviser check about access to rear service road + garage + side passage.

Links to other sections
- *issues for legal adviser (section I)*
- *risks (section J)*

Facilities

Checklist
- shops
- schools
- transport – bus routes, railway station etc.
- local services – doctors, etc.

Please see section H for detailed comments on garages, outbuildings, grounds.

Local shops on main road (busy) 250m away. All usual amenities + buses on main road.

Country park across main road.

Links to other sections
- *issues for legal adviser (section I)*
- *risks (section J)*

Local environment

Checklist
- flood risks
- radon
- EMFs
- mining activity
- contaminated/brownfield sites
- clay sub soils
- coastal erosion
- landslip area
- mundic area
- conservation area

Please see section H for detailed comments on garages, outbuildings, grounds.

Flat – gently sloping site – no flooding problems but check out river in park across main road.

Former mining area but nothing active for decades. Not conservation area. No other problems.

Links to other sections
- *issues for legal adviser (section I)*
- *risks (section J)*

A Surveyor's Guide to the RICS HomeBuyer Service

Property address/Postcode: 52 Recession Ave, Coventry

Site plan/External ....

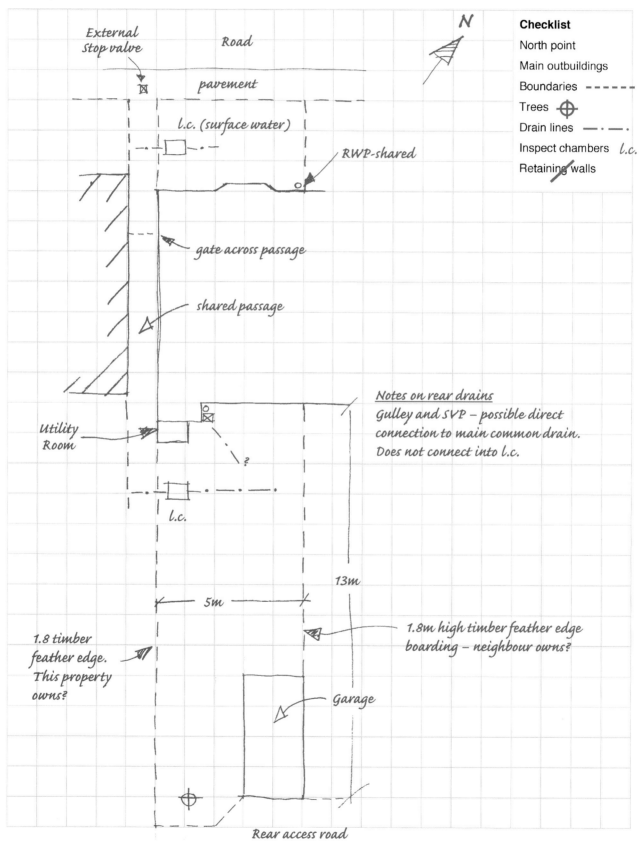

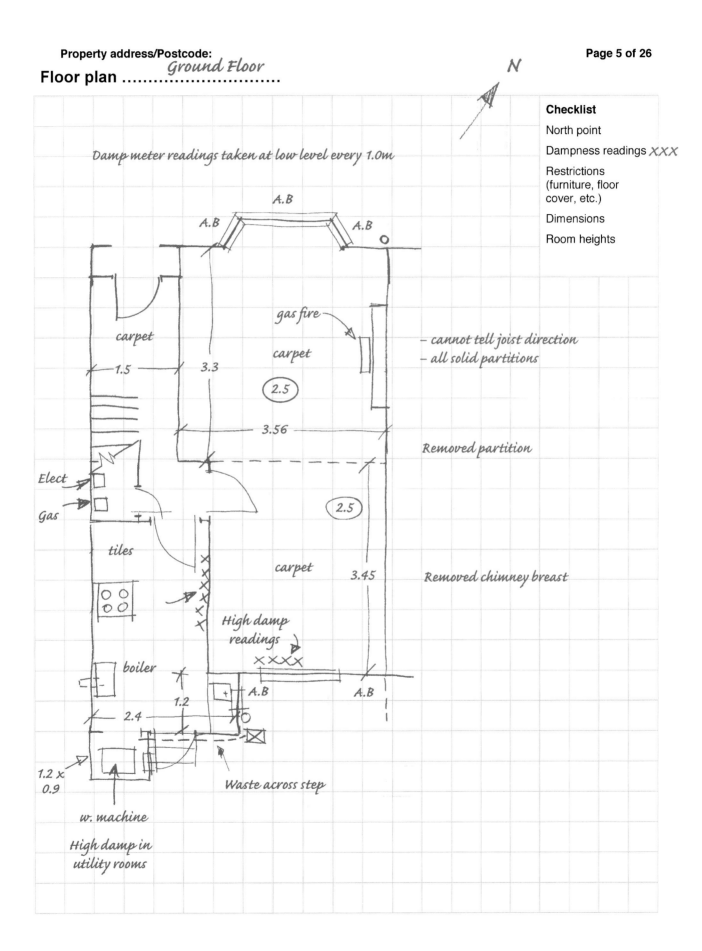

Property address/Postcode:

Floor plan .... *First Floor* ....

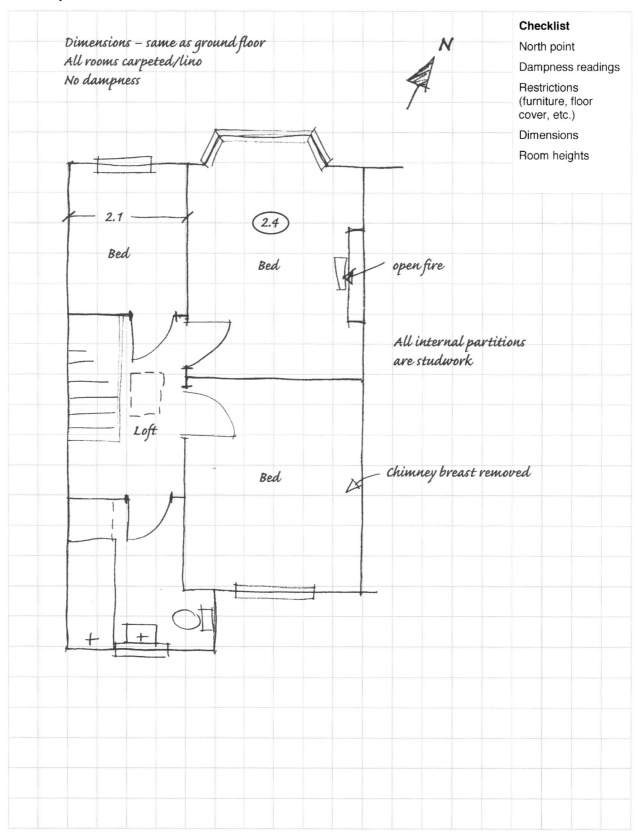

Property address/Postcode:

Section E – Outside the property

E1 Chimney stack	Description of construction and condition
Checklist • description, number and location • flue terminals • chimney pots • flaunching • condition of stacks (leaning, cracked, sulfate attack, etc.) • pointing, render, other finishes • aerials and satellite dishes • flashings and soakers at the junction with the roof covering (but not including the roof covering) • any party wall issues arising from chimney condition • listed building/conservation areas issues **Links to other sections** • cross reference to other affected elements • issues for legal adviser (section I) • risks (section J)	*Front chimney* – brick, two clay pots, no vent on top – open. Shared with next door. TV aerial ok. Flaunching ok. Flashing loose. Pointing ok. *Rear chimney* – similar condition to front but can't properly see pots or flaunch. Flashing loose needs refixing. N.B. no movement to rear despite removal of chimney breast.
E2 Roof coverings	**Description of construction and condition**

Pitched roof
• type of covering and general condition including:
  – the slope/pitch of the roof
  – fixing of tiles, slates
  – temp. repairs
• roofing felt/sarking boarding – presence, type, condition
• ridges and hip tiles
• verge and eaves details
• open valleys
• valley gutters ('butterfly' roofs), parapets gutters, their linings and outlets
• lead flashings to up-stands, dormers, etc. (but not the flashings to the chimney)
• roof lights/roof windows
• dormers (usually including the flat or pitched over the feature)
• any party wall issues arising from the roof condition
• listed building/conservation area issues
 	
Links to other sections
• cross reference to other affected elements
• issues for legal adviser (section I)
• risks (section J)
	Main roof – imitation slate tiles, recently replaced (last 10 yrs) so no asbestos worries. Lead valleys to front gable over bay. Clay ridge + hip tiles. All look ok – bldg reg approval? Next door has concrete tiles, difficult junction between two but ok. Plastic sarking felt beneath – different types but looks ok.

Property address/Postcode:

E2 Roof coverings (cont.)	Description of construction and condition

Checklist
Flat roof
• type of covering and condition including:
  – the falls on the roof
  – pooling of water
  – temp. repairs
  – solar protection
• lead flashings to up-stands (*but not any chimney flashings*)
• other penetrations (*e.g. SVP, vertical flues, etc.*)
• verge and eaves details
• any integral gutters
• roof lights and upstands
• any party wall issues arising from the roof condition;
• listed building/conservation areas
 	
Links to other sections
• cross reference to other affected elements
• issues for legal adviser (section I)
• risks (section J)
	Flat roof (utility)
	Cannot see whole of roof but verge felt damaged, flash band to upstand flashing, no ventilation to structure, but no leaks. Repair but advise for replacement soon.

E3 Rainwater pipes and gutters	**Description of construction and condition**

Checklist
• type of gutters, gutter supports and condition including:
  – falls to gutter
  – position and adequacy of downpipes, hopper, etc.
  – debris/blockage
  – leaks and damage to other parts of building
• swan necks and off-sets
• rainwater downpipes
• hopper heads
• rainwater shoes, etc.
• listed building/conservation areas;
• any party issues with gutters and downpipes, e.g. shared roof drainage, etc.
 	
Links to other sections
• cross reference to other affected elements
• issues for legal adviser (section I)
• risks (section J)
	Plastic RWP + downpipe. Older type but only problem is part missing RWP to rear utility room. Looks like this has never had any downpipe. Rainwater discharging over rear wall – causing damp. Needs new downpipe and connection to drains – can't get into adjacent system.
	
	Rest of RWP + gutter – normal maintenance. Front RWP shared with neighbour – legal checks.

Property address/Postcode:

E4 Main walls	Description of construction and condition

Checklist
• type, thickness and general condition of the wall(s) including:
  – wall ties or sulfate attack
  – wall tie replacement
  – pointing
  – rendering/external finishes
  – cracking/distortion
  – bulging/verticality
  – efflorescence
• cladding:
  – vertical tiles;
  – PVC
  – timber
• damp proof course – type, position, and condition
• evidence of injection holes
• relationship with external ground levels
• presence of dampness to the inside face of the outside wall from rising and penetrating dampness and condensation
• the type and condition of the lintels, jambs and sills
• parapet walls and copings including party up-stand walls
• timber frame walls
• type and extent of insulation to the walls
• any party wall issues arising from the wall condition
• any listed building/conservation area issues
	Walls – solid brick (approx 240-250mm thick) with rough cast render on all elevations. Fair faced brick below window sill to front elevation – mock cavity bond. Fair faced brick soldier course over windows. Blue brick engineering (3 courses) as original dpc. 3 airbricks around front bay – one blocked by plant growth.
	
	Front elevation – satisfactory – no problems.
	
	Side elevation – walls straight, no bulging or cracking, render poorly applied in places and uneven. Brick plinth at low level, blue engineering as before but injected chemical dpc + plastic plugs in offset course. Render poorly made good around balanced flue.
	    render →
	    offset →
	    chem. inject →
	    g.l.
	    engineering
	
	Rear elevation – render poorly applied, especially beneath bathroom window – looks like structural movement but no damage internally – no problem. Render missing around window frames, corners, edges of roof – need repair. Jamb of utility window v. damp.
	
	Dampness – internal face of utility room – disrupted plaster, almost certainly missing downpipe – need replaster once repaired. damp area under lounge window. Chem. dpc as before. Can't see gutter leak or other cause, need to F.I. and check timbers etc.

Links to other sections • cross reference to other affected elements • issues for legal adviser section I) • risks (section J)	Check DPC guarantee Further investigation of dpc/damp problem

Property address/Postcode:

E5 Windows	Description of construction and condition
Checklist • the type and general condition of the windows • orientation of windows • opening of window sample • single/double/secondary glazing • seals to frame • damp penetration around frames • draughts/trickle vents • seals to double glazed units • replacement window issues • escape in the case of fire • low level glass/safety glass • security • listed building/conservation area issues	PVC replacement windows to all openings – relatively modern but no info on spacers. No misting to any panels. Top lights leaded to front, plain to rear. Dampness getting around fixed window to utility room. No trickle vents to any windows. All have escape type windows apart from bathroom, utility and small front and rear bedrooms. Of these, only small front bed a risk. Open front bed, rear lounge + bath. window. All worked ok.
Links to other sections • cross reference to other affected elements • issues for legal adviser (section I) • risks (section J)	Legal check on bldg reg/fensa, if not investigate. J3 – small front bed – window + fire
E6 Outside doors (and patio doors)	Description of construction and condition
Checklist • type/condition of door and frame • level threshold? If level access, exclusion of water, etc. • glazing and safety glass issues, safety glass, etc. • security • draughts/damp penetration around frame • seals to double glazed units • listed building/conservation area issues	PVC double glazed to rear kitchen, original timber glazed + side panels to front. All worked ok, locks working, no damp. Main worry is safety. Front door + side panels – not quite meet benchmark but only by 100mm. Rear – glaze to door + panels clearly below benchmark. No toughened sign on glass. Concerns.
Links to other sections • cross reference to other affected elements • issues for legal adviser (section I) • risks (section J)	Check for back door – bldg reg or fensa – if not the F.I. Warn about front + rear doors J3.

Property address/Postcode:

E7 Conservatories and porches	Description of construction and condition
Checklist • type and condition • inappropriate use as habitable room • single/double glazing • safety - structural stability and safety glazing • snow/debris guards to edge of dwelling roof • separation from rooms of the main dwelling • waterproofing at junction with dwelling • opening of sample of windows • security of windows/doors • effect on escape in the case of fire from the upper windows of the dwelling • guarantee/warranty • building regulation and/or planning approval • listed building/conservation area issues **Links to other sections** • *cross reference to other affected elements* • *issues for legal adviser (section I)* • *risks (section J)*	*Only internal porch to front door – this is ok.*
E8 Other joinery and finishes	**Description of construction and condition**

Checklist
• type and condition of joinery including:
  – *signs of rot*
  – *state of the decorations*
• fascias, soffits and eaves details generally
• bargeboards and verge details;
• ornamental joinery (finials, mock Tudor panelling, etc);
• timber and PVC cladding fixed to main walling
• possible asbestos containing materials
 	
Links to other sections
• *cross reference to other affected elements*
• *issues for legal adviser (section I)*
• *risks (section J)*
	Various timber items including:
	Front elevation – fascia + soffit + mock boarding to gabled roof over bay. Poor decoration, isolated areas of wet to gable area but not serious.
	
	Side elevation – soffit + fascia, condition as above.
	
	Rear elevation – same as front, fascia around the utility flat roof has isolated wet rot. No visual evidence of asbestos products.

Property address/Postcode: Page 12 of 26

E9 Other	Description of construction and condition
Checklist • roof terraces – useable flat roof areas over other accommodation below • balconies • large bay windows where distinctive and different in nature to the construction of the main walls • external stairways and fire escapes, etc. For each feature, include all aspects of the feature, e.g.: • structural aspects – fixings, support, stability, etc. • the walls, floor and roof (if present) to the feature • railings, handrails, etc. **Links to other sections** • *cross reference to other affected elements* • *issues for legal adviser (section I)* • *risks (section J)*	*NA*

Other comments

Include any other external matters here.

Rear chimney breast in loft.

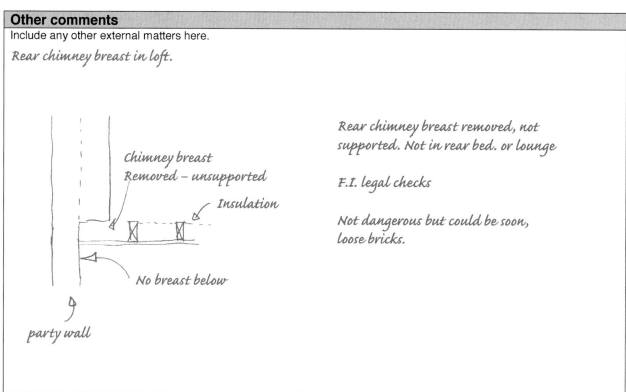

Rear chimney breast removed, not supported. Not in rear bed. or lounge

F.I. legal checks

Not dangerous but could be soon, loose bricks.

Property address/Postcode:

Section F – Inside the property

Restrictions on inspection

Include details of any restrictions on internal inspection:

All rooms had carpets or lino. Could not see under timber floors – no access hatches. Furniture in most rooms but did not prevent inspection. Under stairs cupboard – very packed with possessions. Some possessions in roof.

F1 Roof structure and roof space

Description of construction and condition

Checklist
Pitched roofs
- description of design, construction and condition of the roof structure
- sample moisture readings
- alteration of the roof structure (full attic conversions can be included under F9 Other)
- roof strengthening
- lateral restraint, spreading of roof structure
- the use and misuse of the roof space
- wood rot and wood boring beetle
- party walls – lack of fire breaks, gaps between dwellings
- flooring in loft
- possible asbestos containing materials
- access into loft
- ventilation in roof space
- level of insulation
- Birds, bats, vermin, insects, etc.

Flat roof structure
- access/restrictions to flat roof structure
- sample moisture meter readings
- ventilation to flat roof void
- wood rot and wood boring insect

(Services in the roof space to be included under appropriate elements in section G)

Links to other sections
- cross reference to other affected elements
- issues for legal adviser (section I)
- risks (section J)

Access from landing ok.

Timber roof structure, hipped, double purlin to slopes, usual gable over bay with valleys either side. No distortion or splitting, all joints well made. Could not see dragon ties at corners = possessions + insulation. Checked a few rafters, could not see worm or rot. Binders to ceilings, front + rear. Med moisture readings to underside of back gutters around chimneys. Old staining to some purlins + rafters – sarking felt to roof covering. Impervious plastic, two different types. Covering done out same time, roofer used 2 types. No obvious vent to roof but no signs of any mould/conden. Old doors + wood panels loud loose over joist = dangerous – must be removed.
One-two brick courses missing off top of party wall = gap. Not bad but needs sealing soon. No asbestos seen.

Insulation – looks like loose fill – to back of joists only max. 100mm.

No vermin. No tanks or services in loft.

See page 12 for chimney breast.

Property address/Postcode:

F2 Ceilings	Description of construction and condition
Checklist • description of type and condition including cracking • safety issues – stability of ceiling, bulging, etc. • cornices/centrepieces other decorative features • finishes, including decorations (heavy lining paper, textured coating, etc.) • possible asbestos containing materials, textured decorative coatings, asbestos board, etc. • Listed building features and issues **Links to other sections** • cross reference to other affected elements • issues for legal adviser (section I) • risks (section J)	All original ceilings – part of lath + plaster seen in loft. All ceilings have textured decorative coating – not sure when applied but must include asbestos warning. General unevenness in most room but no serious problems. No staining to upper rooms. Cracking to lounge ceiling where partition removed – not serious but shows there has been some movement. When textured coating removed, expect many repairs.

F3 Walls and ceilings	Description of construction and condition
Checklist • description of type and condition • loading or non-load-bearing partitions • structural movement, cracking, distortion, binding doors, etc. • dampness to internal walls including condensation/mould growth • non-robust partitions, poor sound insulation, low level of fire resistance, etc. • possible asbestos containing material, asbestos board, textured coating, etc. • finishes including decorations • wall tiling (including splash back wall tiling) • dado panelling • structural alterations – through lounge, new door openings, level of support, etc **Links to other sections** • cross reference to other affected elements • issues for legal adviser (section I) • risks (section J)	Ground floor – solid to tap, probably masonry upstairs – all studwork. Bathroom – completely tiled, with tiles painted – higher maint. in future – may damage walls. Kitchen tiling ok. g.f. lounge – central spine wall removed. No evidence of damstand beam, flush ceiling. No nibs to party wall. Movement of door openings to bedrooms above suggest it has deflected. Spirit level used on bedroom floor upstairs – no problems spotted but carpet made it difficult to assess. Legal check + F.I. Not dangerous, but not right. G.F. kitchen – dampness to kitch/rear lounge partition on kitch. side only. Low level could be connected to solid floor at kitch. but not sure. F.I. All walls lined + painted. Don't bother with bldg regs for removed wall – won't have it.

Property address/Postcode:

F4 Floors	Description of construction and condition

Checklist
• description of type and condition (floor by floor)
• solid floor:
  – cracking
  – settlement
  – swelling
• suspended timber floor:
  – stamp test and vibration
  – access hatch?
  – ventilation to sub floor void
• description of floor finishes and their condition
• dampness to all floor types
• wood boring insects
• wood rot
 	
Links to other sections
• cross reference to other affected elements
• issues for legal adviser (section I)
• risks (section J)
	Floors – timber to lounge + hallway: solid to kitchen + utility room. All carpeted or tiled. No access hatches.
	
	Three airbricks + 2 to back. Land falls slightly to back so good height above ground.
	
	No springyness to any floors. Dampness beneath rear lounge window – could affect joist ends. Worn under E4 walls.
	
	On balance, satisfactory.
	
	First floor – generally ok but weakness because of removed spine partition to room below.

F5 Fireplaces, chimney breasts and flues	Description of construction and condition

Checklist
Description of fireplaces, chimney breasts and flues and their condition
• **open, solid fuel fireplaces**:
  – flue/chimney maintenance sweeping
  – lining
  – Part J approvals
  – noncombust. hearth
  – certificates, invoices, etc.
  – fixed external ventilation
  – hearth
• **chimney breasts:**
  – open or blocked off
  – if blocked – ventilation to flue
• terminal at top of flue at chimney level
• condition of chimney breast in roof space:
  – structural alterations
  – unsupported chimney breast
  – dampness to base and face of chimney breasts
	Please note: Actual gas and oil heaters/fires to be included under G4: Heating. This element includes the chimney breast from the lowest level in the house, through all rooms, through any roof space and finishes at the underside of the roof covering. The element changes to 'chimney stack' (E1) above the roof covering where it becomes an external element.
	
	Front chimney breast – g.f. fuel effect gas fire in fireplace. Looks ok but no evidence of maint. agree. No cowl/rainguard on pot. Bedroom fireplace is open fire. Not used but it could be, no hearth carpet taken into fire, dangerous – not to be used. No free vent to room.
	
	Rear chimney breast – removed in roof space. No evidence of support of any type. Bricks loosely cantilevered but not immediately dangerous. No signs of distortion.

Property address/Postcode:

F5 Fireplaces, etc. (cont.)	Description of construction and condition

• independent open vertical flues:
  – description and fixing type, and condition
  – suitability of terminal position
• proximity of flue to combustible materials, etc.
• possible asbestos materials
• listed building issues
 	
Links to other sections
• cross reference to other affected elements
• issues for legal adviser (section I)
• risks (section J)
	Please note: Balanced/fanned flues to be included under G4 Heating. Include here vertical open flues not contained within a chimney.
	
	See previous comments.
	
	Legal adviser check – party wall, breast removed. Maint. checks for gas fires.

F6 Built-in fittings	**Description of construction and condition**

Checklist
Description, adequacy and general condition
 	
• kitchen units, cupboards and worktops:
  – the waterproof sealant but not the splash back tiling (see F3)
• kitchen sinks:
  – the taps (but not the water pipes to them – G3 Water)
  – the slotted wastes (but not the trap or the waste pipe: G6 Drainage)
• dampness affecting the features described above
• wood-boring insects affecting the features described above
• operation and functionality – operable drawers, damaged hinges, etc.
• age and modernity – appeal to purchasers
 	
Links to other sections
• cross reference to other affected elements
• issues for legal adviser (section I)
• risks (section J)
	(Built-in kitchen and other fittings, not including the appliances.)
	
	Kitchen – older style units with decorative timber doors. In satisfactory condition but a few doors need adjusting, drawers overhauling. A little more than normal maintenance.
	
	No other fitted cupboards – all standing furniture.

Other comments

Property address/Postcode:

F7 Woodwork	Description of construction and condition
Checklist Description and general condition: • internal doors, frames/linings, architraves, etc. • internal parts of windows – jamb linings, sill boards, architraves, shutters, etc. • built in cupboards – for example within chimney alcoves, meter cupboards, understairs • skirtings, dado rails, picture rails • staircases – treads and risers, strings, balusters, handrails, newel posts: – safety hazards – missing balusters, broken treads, etc. • finishes to features above including decorations • dampness affecting the features described above • wood rot and wood-boring beetle affecting the features described above **Links to other sections** • cross reference to other affected elements • issues for legal adviser (section I) • risks (section J)	Entrance area – panelled door + partition to understairs cupboard, door works ok, no dampness but difficult to get right under because of possessions. Balusters + handrail missing = safety hazard. Stairs fully carpeted but feel ok – no creaks. No woodworm on underside. Internal doors – g.f. newer Mdf types, panelled + ok. F.f. – original panelled + one hardboard door. Ok but a few need latches adjusting – norm. maintenance. Doors to front + rear bedrooms binding because of removed int. partition. Skirtings ok – but those in utility + rear lounge may need repairing because of dampness. Bathroom cupboard – poor quality but functioning. J3 – lack of handrails.

F8 Bathroom fittings	Description of construction and condition
Checklist Description, adequacy and general condition: • baths, bidets, wash hand basins, their taps and slotted wastes but **not** the water pipes to them (G3 and G5), and **not** the traps and waste pipe (G6) • shower trays, shower cubicles but **not** the mixer valve (G3) or electric shower (G1) • WCs, their cisterns and overflows. To include junction between the WC outgo and the drainage pipe but not the drainage pipe itself (G6) • bath panels and other built-in fittings associated with the sanitary appliances • the sealant between the appliance and the adjacent surfaces but not the splash back tiles (see F3) • leaks and dampness caused by the bathroom fittings • wood rot and wood-boring insects affecting the features described above **Links to other sections** • cross reference to other affected elements • issues for legal adviser (section I) • risks (section J)	One bathroom – low level flush WC – wash hand basin – bath with electric shower over + shower screen All functioning, no leaks, renew water proof seals around fittings. Older types, need continued maintenance. Warn of this.

Property address/Postcode:

F9 Other	Description of construction and condition
Checklist This section is for anything internal that does not sit comfortably in earlier sections in section F. This could typically include: • **Roof space conversion** – include here where a total conversion has been carried out. To give a complete appraisal of the conversion, include all elements here. For example: – *structural aspects – alteration of roof, strengthening of ceilings and other floors* – *access to room – adequacy and safety of staircase, etc.* – *thermal insulation – including ventilation to roof spaces* – *fire escape route – fire resistance of partitions, ceilings, doors, etc.* – *regulation and permissions – building regulation, planning permission, listed building/conservation, etc.* • **Cellar/basements** – where the cellar/basements are used for habitable purposes, to give a complete appraisal, include all elements here. For example: – *structural aspects – removal of partitions, strengthening of ceilings and floors* – *access to room – adequacy and safety of staircase, etc.* – *fire escape and fire safety – additional fire resistance to ceiling, inner room issues, etc.* – *daylight and ventilation to rooms and spaces – compare with building regs* – *dampness – tanking issues, guarantees, etc.* • **Flats** – to include those internal common parts of a block: – *internal hallways, landings and stairs* – *doors and windows* – *built-in cupboards, meter cupboards, etc.* – *asbestos containing materials* – *asbestos management plan* **Links to other sections** • *cross reference to other affected elements* • *issues for legal adviser (section I)* • *risks (section J)*	NA

Property address/Postcode:

Section G - services
Restrictions on inspection
Include details of any restrictions on the inspection of the services: *Vacant property – no info from agent, services switched off.*

G1 Electricity	Description of construction and condition
Checklist • description and general condition • evidence of installation, alteration or maintenance certification (e.g. Part P approval, minor works cert., periodic inspections and reports, etc. **Main issues** (broad indicators only): • is there a mains supply • on-peak/off-peak supply • location of the meter and consumer unit/fuse board • separate from gas meter • is there a RCD/MCB • equipotential bonding in the usual places • type and condition of visible wiring • condition of a sample of the range of light fittings and switch gear • type and condition of fixed electrical appliances including heaters, storage radiators, electric showers, instant water heaters, etc. (see also G5 water heating) • bath and shower rooms – nature and suitability of electrical fittings • external installations – garages, outbuildings, external sockets, garden lighting, water feature pumps, etc. • external installations – covered by RCD **Links to other sections** • cross reference to other affected elements • issues for legal adviser (section I) • risks (section J)	The following general advice will be printed in every report: Safety warning: Periodic inspection and testing of electrical installations is important to protect your home from damage and to ensure the safety of the occupants. Guidance published by the Institution of Electrical Engineers recommends that inspections and testing are undertaken at least every ten years and on change of occupancy. All electrical installation work undertaken after 1 January 2005 should be identified by an Electrical Installation Certificate. *Mains electric with meter + consumer unit under stairs. Relatively recent board but not MCBs or RCD. Old light switch + cable in understairs. No evidence of last test. Evidence of supp. bond to gas pipe.* *All switches + sockets modern + pendants + light fittings. Not enough sockets. Halogen lights in bathroom = hazard.* *Unusual cable to outside light behind wash hand basin.* *Not much supp. bonding in bathroom.* [*Supp. bond on rising main*] *External light above utility room.* *Garage? Could not get in so not sure if the garage has any.* *J3 – lack of records bath lights older switch.*

Property address/Postcode:

G2 Gas/Oil	Description of construction and condition

Please note: this section covers the gas and oil supply pipework from the utility company's supply (or the storage vessel) through to the heating appliance itself. All matters relating to the appliance are to be reported under G4 Heating.
 	
Checklist
• description and general condition
• evidence of installation, alteration or maintenance certification (e.g. Gas Safe Registered engineer reports, OFTEC certs., service agreement, etc.)
• **mains gas installations:**
• does it have a mains supply and is it connected
• meter location and condition – is it positioned on an escape route
• the route, nature and condition of the pipework – well fixed and supported
• have there been alterations and are they properly certified
• **LPG installations:**
• are the storage cylinders in a suitable location:
  – boundaries
  – proximity to buildings (especially basements), other storage tanks
• condition of tank, covered with foliage, etc
• the route, nature and condition of the pipework
• have there been alterations, and are they properly certified?
• **oil installations:**
• is this satisfactory, including:
• the position of tank:
  – proximity to dwelling, boundaries, watercourses, drains, etc.
  – seen from filling point
• nature of the tank:
  – internal or external bunded
  – condition of tank
 	
Links to other sections
• cross reference to other affected elements
• issues for legal adviser (section I)
• risks (section J)
	The following general advice will be printed in every report:
	Safety Warning: Regular inspection, testing, maintenance and servicing of all gas/oil appliances and equipment should be undertaken by a registered 'competent person' and in accordance with the manufacturer's instructions. This is important to ensure that such equipment is working correctly to minimise the risk of fire and carbon monoxide poisoning as well as leakages of carbon dioxide and other greenhouse gases to the atmosphere. For further advice contact Gas Safe Register for gas installations, and OFTEC for oil installations.
	
	Mains gas-meter under stairs. Not fixed properly – can be moved. Heavy possessions piled against it. Pipe dented in places, lead gas pipes. No evidence of test or inspection. Potential hazard.
	
	J3 Gas meter → gas leaks

Other comments

Property address/Postcode:

G3 Water	Description of construction and condition

Checklist
General description of the system and condition including:
• the piped water supply from the boundary of the property:
  – pipe material (lead?)
  – external stop valve
  – depth of pipe?
  – water meter
  – where does it enter building?
• Internally, where the rising main appears:
  – stop valve – accessibility
  – insulation (especially underfloor and in roof space)
  – material – lead? dissimilar materials?
  – support to pipework
• cold water storage and expansion tanks:
  – support to tank
  – tank type and condition
  – lid, thermal insulation
  – overflow
• backflow prevention to:
  – bidet, etc.
  – garden hose pipe,
  – irrigation system
 	
Private water supplies
Main issues to include
• description and general condition
• private supply only or combined with mains
• nature of source
• has supply been checked and certified as wholesome source of supply
• shared supply with neighbours
 	
Links to other sections
• cross reference to other affected elements
• issues for legal adviser (section I)
• risks (section J)
	Please note: This element covers cold water supply service to combination boilers, water heaters, hot water cylinders but not the pipe work beyond them (see G5). It covers cold water storage tanks and the down service beyond, but **not** the taps of appliances.
	
	Mains water stop valve in pavement at top of passage. 2 stop valves so assume one for each property. No meter.
	
	Can't see much of pipework in copper where visible – no storage tanks so all appliances served directly off the main. Main rises beneath sink in kitch. in copper with stop valve but no drain-down visible. Likely main rises through solid floor. No outside tap. No lead visible.
	
	In bathroom mixer valve to shower connection to h + c supplies – looks ok but not water on.

Other comments

Property address/Postcode:

G4 Heating	Description of construction and condition

Checklist
General description and condition of the heating system(s):
• evidence of certification for installation, maintenance, repair etc. (see also G2)
• boilers (including combination boilers that also provide hot water)
• gas and oil fires and heaters
• electric heating (but not the electricity supply to them
• heat pumps
• other forms (describe)
• heat distribution pipework
• heat emitters – radiators, panel heaters, under-floor heating pipework, etc.
• ancillary equipment including:
  – heating controls
  – expansion vessels and safety controls
• ventilation issues for heating appliances (where appropriate)
Please note: free standing solid fuel heating sources (incl. Aga) should be covered here. Where in fireplace include in F5 but distribution pipework, etc. covered under this element.
 	
Links to other sections
• cross reference to other affected elements
• issues for legal adviser (section I)
• risks (section J)
	Please note: Most heating appliances will be subject to regulations governing installation of gas, oil, solid fuel and electric appliances previously described. Cross reference to appropriate section where certification is required.
	
	Condensing combi boiler in kitchen – wall mounted balanced flue – looks like replaced previous. Pressure relief pipe ok but conden. pipe discharge over ground? Looks ok – modern (Greenstar 24i) programmer on the boiler, room thermostat in hall, no TRVs – manual hand wheel types rads in every space. Not vent required.
	
	Damp near to rad in kitch. but I don't think rad is the cause.
	
	Check it regularly maintenance. Bldg regs or competent person?
	
	J3 absence of maint. records.

G5 Water heating	Description of construction and condition
Checklist Include the hot water heating appliance where it provides hot water only, for example: • gas water heater • multi-point water heater • electric under-counter heater • instantaneous electric water heater; • solar water/PV heating Other parts of hot water system to include: • the hot water cylinder • insulation to tank and primary pipe work • expansion pipe • immersion heaters • cylinder thermostats • all distribution pipework between cylinder and the appliances (but not the taps) • the distribution pipework between water heating source and the various appliances **Links to other sections** • issues for legal adviser (section I) • risks (section J)	**Please note:** Where hot water provided by combination heating, hot water boilers, heat pumps, etc. include under G4. Evidence of certification if not covered in G2/G4. *All hot water provided by combi via copper pipes where visible. Shower mixer also has hot water supply.* *Comments as above.*

Property address/Postcode:

G6 Drainage	Description of construction and condition

Checklist
General description and condition of drainage system.
Above ground drainage:
• traps and wastes from sinks, washing machines, dishwashers, baths, bidets, WCs, showers, wash hand basin, etc.
• pipe material, condition, leaks
• length of wastes, cleaning access
• slope of branch connections
• soil and vent pipes, two pipe systems, etc.
• proximity of stack to windows
• stub stacks
• air admittance valves, etc.
Below ground drainage
• gullies – type, condition, grid, blocked
• clearing access to drains
• inspection chambers:
  – position
  – cover/condition
  – walls/benching/channel
  – standard of connections
  – Bldg regs for new work
  – interceptor
  – blocked
• underground drains:
  – type
  – building over drains
  – trees close by?
  – depth and approx fall?
• design of system:
  – separate soil and surface
  – combined system
  – inappropriate connections
  – join with other properties – private sewer?
Private drainage including:
• design and type:
  – cess pit
  – septic tank (two chamber)
  – small sewerage treatment system (e.g. Klargester, etc)
• Other types:
  – soakaway
  – reed beds
  – composting toilets.
• Evidence of maintenance, permission to discharge, etc.
• Shared system
• Approx. size? Adequacy
• Evidence of misuse
  – RWP connection
  – Extra WCs to small tank
 	
Links to other sections
• cross reference to other affected elements
• issues for legal adviser (section I)
• risks (section J)
	Above ground drainage
	
	Bathroom
	
	WC – outgo directly into PVC soil + vent. Bath + handbasin into same waste and discharge over hopper head externally.
	
	Kitchen
	
	PVC waste directly over ext. gulley. Waste from wash machine in utility room over rear step and into gulley at base of SVP. Tripping hazard.
	
	Head of SVP too close to window head = smells. At base, PVC SVP connected into older section of cast iron. No rodding eye and because no connection to I.C. then problem if blocked.
	
	Below ground drainage
	
	Front
	
	I.C. + drain across front, serves at least 3 houses. Probably surface water only. Cracked cover = hazard. Brick walls poor, benching almost gone, channel full of pebbles (next door garden + blocked) but no standing water. Debris suggest pipe broken somewhere. Need clearing + F.I. Best re-build whole I.C.
	
	Rear
	
	I.C. in rear garden – broken cover + poor condition – not as bad as front I.C. No direct connection from SVP + adjacent gulley. This means difficult to clear if blocked. Large shrub growing right next to gulley.
	
	Shared drain front + back.
	
	Missing RWP from utility roof + trip hazard waste both need new connection to I.C. in back garden.
	
	J3 – Utility waste tripping hazard.

Property address/Postcode:

G7 Common services	Description of construction and condition
Checklist General description and condition of the facilities to the block General comments in relation to: • refuse chutes, bin stores, etc. • lifts – passenger and goods • entry-phone systems • CCTV • common areas – heating and lighting • communal heat and power • renewable energy systems **Links to other sections** • *cross reference to other affected elements* • *issues for legal adviser (section I)* • *risks (section J)*	NA

Other comments

Appendices

Property address/Postcode:

Section H – Grounds (including shared areas for flats)
Restrictions on inspection
Include details of any restrictions on the inspection of the services: *Rear garden is very overgrown, can't see part of garage. Garage locked and window panel dirty – could not see in.*

H1 Garage	**Description of construction and condition**

Checklist
General description, location, and condition including all elements and services, including:
• functionality:
  – size for modern cars
  – access to entrance
• condition:
  – structure
  – walls/cladding
  – floor (inspection pits?)
  – roof – leaks, fragile covering
  – asbestos containing materials
  – windows/doors
• services:
  – electrics (RCD protection?)
  – heating
  – water
  – drainage
• unauthorised use – commercial use, office, habitable, etc.
• conservation area, listed buildings issues
Please note: Integral garages should be including in the main report (mostly section F) under the elemental rating system.
 	
Links to other sections
• issues for legal adviser (section I)
• risks (section J)
	Pre-cast sectionalised concrete garage to rear of rear garden. Metal up + over door. Access to garage only from rear access road and this is gated + locked near to where access road connection to acacia avenue. Therefore not used.
	
	Roof at garage is asbestos cement, covered with foliage so can't see condition – give usual warning. Also fragile cement sheets. Can't see any services but don't know.
	
	Check access road + maint. responsibilities.
	J3 Fragile roof, potential asbestos

H2 Other	**Description of construction and condition**

Checklist
Other permanent outbuildings:
• summer houses
• permanent greenhouses that are to remain
• follies
• substantial garden sheds that are to remain
• leisure buildings (but not the leisure facilities inside)
• for each of the above:
  – general description and condition including roof, walls, floors, windows, doors, etc.
  – services
  – electrics (RCD protection?)
  – heating
  – water
  – drainage
• unauthorised use – commercial use, office, habitable, etc.
• conservation area, listed building issues
 	
Links to other sections
• issues for legal adviser (section I)
• risks (section J)
	No other outbldgs – only garage.

Property address/Postcode:

H3 General	Description of construction and condition
Checklist To include a general description of the grounds and/or site including: • boundaries – condition and likely ownership, changed positions, etc. • retaining walls – construction, type, condition, etc. • terraces and patios • conservation and listed building issues • gardens – general description, approx. size, landscape/garden design features • drives and paths – type and condition • steps and changes of level – type, construction safety • trees – type, location to buildings and drains, maintenance, safety • ponds, lakes, reed beds, drainage fields, water courses, etc. – type, size, condition, ownership implications, safety • rights of way/easements across site/wayleaves, etc. • wider location/environment issues where impinge within boundaries, e.g. contamination, HV pylons, HV sub stations, landslip, etc. **Links to other sections** • issues for legal adviser (section I) • risks (section J)	*Boundaries* *Rear* – all feather edge on short conc. posts. SW under ownership at this property? Several arris rails are rotten + repairs required. Not replacement at this stage. *Front* – stone wall (dwarf) at front, side of entrance is collapsed – need repair. To SW side, timber railed + timber post – poor condition. No fence between front gardens. *Footpaths* – front to entrance door – conc. + tripping hazards, side path = gravel = ok. Rear garden. Informal paved patio – poor condition + trip hazard. Disused B-B-Q construction. *Rear garden* – part lawn but v. overgrown with large shrubs and one tree to rear. Front garden – rough lawn – lylandii behind front garden wall causing it to collapse. Reduce size of trees in front garden or removed. Same to tree near garage in rear garden. Check ownership of fences Front paths – tripping hazards

Other comments

APPENDIX 5 *Complete report (sample)*

RICS HomeBuyer REPORT...

Property survey and valuation

Date of inspection:	25/12/2009
Property address:	52 Recession Avenue, Coventry, Warwickshire, CV0 8JJ
Customer's name:	Mr A.N. Other

A Surveyor's Guide to the RICS HomeBuyer Service

RICS
**Home
Buyer**
REPORT...

Contents

A   Introduction to the report
B   About the inspection
C   Summary of the condition ratings
D   About the property
E   Outside of the property
F   Inside of the property
G   Services
H   Grounds (including shared areas for flats)
I   Issues for your legal advisers
J   Risks
K   Valuation
L   Surveyor's declaration
     What to do now
     Description of the HomeBuyer Service

RICS is the world's leading qualification when it comes to professional standards in land, property and construction.

In a world where more and more people, governments, banks and commercial organisations demand greater certainty of professional standards and ethics, attaining RICS status is the recognised mark of property professionalism.

Over 100,000 property professionals working in the major established and emerging economies of the world have already recognised the importance of securing RICS status by becoming members.

RICS is an independent professional body originally established in the UK by Royal Charter. Since 1868, RICS has been committed to setting and upholding the highest standards of excellence and integrity - providing impartial, authoritative advice on key issues affecting businesses and society.

Please read the 'Description of the RICS HomeBuyer Service' (inside back cover) for details of what is, or is not, inspected.

© RICS 2009

the mark of property professionalism worldwide

Property Address: 52 Recession Avenue, Coventry, Warwickshire, CV0 8JJ
Serial Number: 43f6202a-7431-45a0-a819-c86a909ead96

A Introduction to the report

RICS Home Buyer REPORT...

This HomeBuyer Report is produced by an RICS surveyor ('the surveyor'). They have written this report for you to use. If you decide not to act on the advice in this report, you do this at your own risk.

The HomeBuyer report aims to help you:

- make a reasoned and informed decision on whether to go ahead with buying the property;

- make an informed decision on what is a reasonable price to pay for the property;

- take account of any repairs or replacements the property needs; and

- consider what further advice you should take before exchanging contracts (if the property is in England, Wales, Northern Ireland, the Isle of Man or the Channel Islands), or before concluding an offer (if the property is in Scotland).

Any extra services the surveyor provides are not covered by these terms and conditions and must be covered by a separate contract.

If you want to complain about the service, please refer to the complaints handling procedure in the 'Description of the RICS HomeBuyer Service' enclosed with this report.

Please read the 'Description of the RICS HomeBuyer Service' (at the back of this report) for full details of what is, and is not, inspected.

Please read the 'Description of the RICS Homebuyer Service' (inside back cover) for details of what is, or is not, inspected.

© RICS 2009

Property Address: 52 Recession Avenue, Coventry, Warwickshire, CV0 8JJ
Serial Number: 43f6202a-7431-45a0-a819-c86a909ead96

RICS — the mark of property professionalism worldwide

About the inspection

RICS **Home Buyer** REPORT...

Surveyor's name: Phil Parnham

Surveyor's RICS number: 007

Company name: BlueBox partners

Date of the inspection: 25/12/2009

Report reference number: 00001

Related party disclosure:

I have no links to this transaction.

Full address and postcode of the property:

52 Recession Avenue, Coventry, Warwickshire, CV0 8JJ

Weather conditions when the inspection took place:

When I inspected the home, the weather was dry with sunny periods following a period of dry weather.

The status of the property when the inspection took place:

When I inspected the property, it was unoccupied; partly furnished and the floors were fully covered.

Please read the 'Description of the RICS HomeBuyer Service' (inside back cover) for details of what is, or is not, inspected.

© RICS 2009

the mark of property professionalism worldwide

Property Address: 52 Recession Avenue, Coventry, Warwickshire, CV0 8JJ
Serial Number: 43f6202a-7431-45a0-a819-c86a909ead96

B About the inspection (continued)

The surveyor's inspection

RICS
Home Buyer
REPORT...

The surveyor inspects the inside and outside of the main building and all permanent outbuildings, but does not force or open up the fabric. They also inspect the parts of the electricity, gas/oil, water, heating and drainage services that can be seen, but they do not test them.

To help describe the condition of the home, the surveyor gives condition ratings to the main parts (or 'elements') of the building, garage, and some parts outside. Some elements can be made up of several different parts. For example, a home can have a pitched (sloped) roof to the main building and a flat roof to an extension. In this case, the surveyor will give each a condition rating and use the worst one to define the element in section C.

In the element boxes in parts E, F, G and H, the surveyor will describe the part that has the worst condition rating first and then briefly outline the condition of the other parts. The condition ratings are described as follows.

Important note

The surveyor will carry out only a visual inspection. This means that they do not take up carpets, floor coverings or floorboards, move furniture or remove the contents of cupboards. Also, they do not remove secured panels or undo electrical fittings.

The surveyor will inspect roofs, chimneys and other surfaces on the outside of the building from ground level, and, if necessary, from neighbouring public property and with the help of binoculars.

They will inspect the roof structure from inside the roof space if there is access (although they will not move or lift insulation material, stored goods or other contents). They will examine floor surfaces and under-floor spaces so far as there is access to these (although the surveyor will not move or lift furniture, floor coverings or other contents). The surveyor is not able to assess the condition of the inside of any chimney, boiler or other flues.

The surveyor will note in their report if they were not able to check any parts of the property that the inspection would normally cover. If the surveyor is concerned about these parts, the report tells you about any further investigations that are needed.

The surveyor will not report on the cost of any work to put right defects or make recommendations on how these repairs should be carried out. Some maintenance and repairs the surveyor suggests may be expensive.

Condition rating	Definition
	Defects which are serious and/or need to be repaired, replaced or investigated urgently.
	Defects that need repairing or replacing but are not considered to be either serious or urgent. The property must be maintained in the normal way.
	No repair is currently needed. The property must be maintained in the normal way.
	Not inspected. (See 'Important note' in the panel opposite.)

The report focuses on matters that, in the surveyor's opinion, need to be dealt with or may affect the value of the property.

 Please read the 'Description of the RICS Homebuyer Service' (inside back cover) for details of what is, or is not, inspected.

© RICS 2009

Property Address: 52 Recession Avenue, Coventry, Warwickshire, CV0 8JJ
Serial Number: 43f6202a-7431-45a0-a819-c86a909ead96

 the mark of property professionalism worldwide

A Surveyor's Guide to the RICS HomeBuyer Service

Summary of the condition ratings

RICS
Home
Buyer
REPORT...

This section summarises the condition ratings of the different elements of the property.

If an element is made up of a number of different parts (for example, a pitched roof to the main building and a flat roof to an extension), only the part in the worst condition is shown here.

To make sure you get a balanced impression of the property, we strongly recommend that you read all sections of the report.

3   **Condition rating 3:** Defects which are serious and /or need to be repaired, replaced or investigated urgently.

Section of the report:	Element number:	Element name:
E: Outside of the property	E3	Rainwater pipes and gutter
	E4	Main walls
F: Inside of the property	F1	Roof structure
	F3	Walls and partitions
	F5	Fireplaces, chimney breasts and flues
	F7	Woodwork (for example staircase and joinery)
G: Services	G1	Electricity
	G2	Gas/oil
	G4	Heating
	G5	Water heating
	G6	Drainage

 Please read the 'Description of the RICS HomeBuyer Service' (inside back cover) for details of what is, or is not, inspected.

© RICS 2009

the mark of property professionalism worldwide

Property Address: 52 Recession Avenue, Coventry, Warwickshire, CV0 8JJ
Serial Number: 43f6202a-7431-45a0-a819-c86a909ead96

Condition rating 2: Defects that need repairing or replacing but are not considered to be either serious or urgent. The property must be maintained in the normal way.

Section of the report:	Element number:	Element name:
E: Outside of the property	E1	Chimney stacks
	E2	Roof coverings
	E8	Other joinery and finishes
F: Inside of the property	F2	Ceilings
	F6	Built-in fittings (eg. built in kitchen, not appliances)
	F8	Bathroom fittings

Condition rating 1: No repair is currently needed. The property must be maintained in the normal way.

Section of the report:	Element number:	Element name:
E: Outside of the property	E5	Windows
	E6	Outside doors (including patio doors)
	E7	Conservatory and porches
F: Inside of the property	F4	Floors
G: Services	G3	Water

Please read the 'Description of the RICS Homebuyer Service' (inside back cover) for details of what is, or is not, inspected.

© RICS 2009

Property Address: 52 Recession Avenue, Coventry, Warwickshire, CV0 8JJ
Serial Number: 43f6202a-7431-45a0-a819-c86a909ead96

the mark of property professionalism worldwide

C Summary of the condition ratings (continued)

The surveyor's overall opinion of the property:

This property is considered to be a reasonable proposition for purchase at a price of £135 000, provided that you are prepared to accept the cost and inconvenience of dealing with the various repair/improvement works reported. These deficiencies are common in properties of this age and type. Provided that the necessary works are carried out to a satisfactory standard, I see no reason why there should be any special difficulty on resale in normal market conditions.

 Please read the 'Description of the RICS HomeBuyer Service' (inside back cover) for details of what is, or is not, inspected.

© RICS 2009

Property Address: 52 Recession Avenue, Coventry, Warwickshire, CV0 8JJ
Serial Number: 43f6202a-7431-45a0-a819-c86a909ead96

 the mark of property professionalism worldwide

Appendices

D About the property

RICS Home Buyer REPORT...

Type of property: The home is an end of terrace house and has two storeys. The front of the home faces North West.

Approximate year the property was built: Based on my knowledge of the area and housing styles, I think the home was built between 1930 - 1935.

Approximate year the property was extended: A utility room extension has been added to the home and based on my knowledge of construction, I think this was built in 1960 - 65.

Approximate year the property was converted: Not applicable.

For flats and maisonettes: Not applicable.

Accommodation

Floor	Living rooms	Bed-rooms	Bath or shower	Separate toilet	Kitchen	Utility room	Conser-vatory	Other	Name of other
Lower ground									
Ground	2				1	1			
First		3	1						
Second									
Third									
Other									
Roof space									

Construction:

The home is constructed using traditional materials and techniques.
The roof is constructed of timber and covered with tiles.
The upper floors are made of timber.
The ground floor is made of a mixture of timber and solid concrete.
The outside walls are made of brick with a cement coating called render.

Please read the 'Description of the RICS Homebuyer Service' (inside back cover) for details of what is, or is not, inspected.

© RICS 2009

Property Address: 52 Recession Avenue, Coventry, Warwickshire, CV0 8JJ
Serial Number: 43f6202a-7431-45a0-a819-c86a909ead96

the mark of property professionalism worldwide

D About the property (continued)

Energy

The property's energy performance, as recorded in the Energy Performance Certificate

■ Energy-efficiency rating — The Energy Performance Certificate was not available.

■ Environmental impact rating — The Energy Performance Certificate was not available.

Outside the property

The home has the following gardens: to the front and rear and there is a shared passageway that connects to the access roadway at the rear. There is no on-site parking. Off-site parking is not restricted.
There is a single garage in the rear of the rear garden of the property. Apart from this, there are no outbuildings.

The location

The home is in a well-established residential area one and a half kilometres North West of Coventry city centre.

Facilities

The local facilities include: shops and other retail outlets on the main road approximately 250 metres away and local bus routes and schools.

Local environment

The home is in an area that is unlikely to flood.
The home is in an area that has historically been affected by mining and indictrial activity but this does not pose any special risks.

RICS Home Buyer REPORT...

Mains services

Drainage ✓   Gas ✓   Electricity ✓
Water ✓

The ticked boxes show that the mains services are present.

Central heating

Electric ☐   Gas ✓   Solid fuel ☐
Oil ☐

Other services or energy sources:

Not applicable

Please read the 'Description of the RICS HomeBuyer Service' (inside back cover) for details of what is, or is not, inspected.

© RICS 2009

Property Address: 52 Recession Avenue, Coventry, Warwickshire, CV0 8JJ
Serial Number: 43f6202a-7431-45a0-a819-c86a909ead96

the mark of property professionalism worldwide

Appendices

Outside of the property

RICS Home Buyer REPORT...

I could only partially inspect the rear chimney stack because the rear garden was overgrown and restricted my inspection.

Condition rating

E1 Chimney stacks

The two chimney stacks to the front and rear are made of brick and the front chimney is fitted with clay chimney pots. The waterproofing between the chimney stack and the roof covering (called the flashing) is loose. **Condition rating 2.**

The roof covering and chimney stack may have to be disturbed to repair the flashings and to repair the chimney safely, contractors will have to use appropriate access equipment (e.g. scaffolding, hydraulic platforms, etc.). Both of these can increase the cost of the work (see F5 for capping the front chimney flues and the partially removed chimney breast to the rear).

E2 Roof coverings

Flat roof
The flat roof over the rear utility room is covered with roofing felt. The roof covering around the edge is damaged. **Condition rating 2.**

Flat roof coverings do not last as long as those on sloping roofs and they can quickly get worse. Although the damaged areas can be replaced, you may find it economic to replace the whole roof covering now.

Main roof
The main sloping roof is covered with imitation slates over a secondary waterproof barrier (roofing felt). **Condition rating 1.**

The roof covering is relatively new and you should ask your legal adviser if this work has Building Regulation approval from the local authority (see section I1). If not, you should ask an appropriately qualified person to inspect the roof and check its stability.

 Please read the 'Description of the RICS Homebuyer Service' (inside back cover) for details of what is, or is not, inspected.

© RICS 2009

Property Address: 52 Recession Avenue, Coventry, Warwickshire, CV0 8JJ
Serial Number: 43f6202a-7431-45a0-a819-c86a909ead96

 the mark of property professionalism worldwide

E Outside of the property (continued)

E3 Rainwater pipes and gutter

Rainwater pipes and gutters to the utility room
The pipe to the utility room is incomplete. This is causing dampness in the walls of the utility room (see E4). **Condition rating 3.**

The lack of a rainwater pipe to the Utility room is causing dampness to the building below and you should provide one now. When this is done, you will have to provide a new connection to the underground drainage system (see section G6). This work may be costly.

Rainwater pipes and gutters to the main house
The rainwater pipes and gutters to the main house are made of plastic. **Condition rating 1.**
The rainwater pipe to the front of the house is shared with the neighbouring property (No.50 Recession Avenue). You should check with your legal adviser about your rights and obligations (see section I3).

E4 Main walls

Walls to the main house
The external walls are made of brick. The outer faces of the wall are covered by a coating of cement (render). The outside walls have a barrier against dampness rising from the ground (called a damp-proof course or DPC) consisting of two rows of dense blue bricks and a subsequent chemical dpc has been injected into the wall. There are a number of defects:

- Dampness affects the inside face of the outside wall at low level in the rear lounge but I could not discover the cause (see section J1);
- the render is damaged and missing in a number of places (for example, around the window openings, junctions with the roof covering and around the rainwater fittings).

Condition rating 3 (further investigation).

The dampness may be causing problems in other parts of the building and so you should:

- ask your legal adviser to check whether there is a valid guarantee or warranty that covers this work (see section I), and;
- if not, you should ask an appropriately qualified person to inspect the problem and provide you with a report. To do this properly, parts of the property will have to be disturbed and you should discuss this with the current owner. This investigation should also include the dampness in the kitchen wall (see F3).

Once you have repaired the wall, you may have to carry out other additional work and this could typically include replacing damp plaster, repairing rotten timber, etc. This can add to the cost of the work.

Property Address: 52 Recession Avenue, Coventry, Warwickshire, CV0 8JJ
Serial Number: 43f6202a-7431-45a0-a819-c86a909ead96

E Outside of the property (continued)

E4 Main walls

Walls to the utility room extension
The external walls are made of brick. The outer faces of the wall are covered by a coating of cement (render). The inside of all the walls are damp because of the missing rainwater pipe and gaps around window frames. **Condition rating 3.**

When you have done the repairs you will have to carry out other work. This typically includes repairing the render, replacing damaged internal plaster, and checking and repairing any rotten timber components. This can add to the cost of the work.
Render coating beneath the bathroom window
The render coating below the first floor bathroom window is uneven because of poor workmanship but this is not significant.

E5 Windows

The windows are replacement double glazing in PVC frames. **Condition rating 1.**

The design of the small front bedroom window does not allow people to escape from a fire and this is a safety hazard. In a fire, smoke can quickly trap occupants in a room and the small size of the opening will prevent them escaping or others rescuing them. To reduce this hazard, you should ask an appropriately qualified person for advice (see Section J3).

You should ask your legal adviser to check whether these windows have either Building Regulation approval or have been installed by a contractor registered with FENSA (see section I1). This is a government-approved trade association whose members can self-certify that their installations meet the standards of the Building Regulations. If they were installed before April 2002 or do not have either of these, you should ask an appropriately qualified person to assess the quality of the installation.

Please read the 'Description of the RICS Homebuyer Service' (inside back cover) for details of what is, or is not, inspected.

© RICS 2009

Property Address: 52 Recession Avenue, Coventry, Warwickshire, CV0 8JJ
Serial Number: 43f6202a-7431-45a0-a819-c86a909ead96

the mark of property professionalism worldwide

A Surveyor's Guide to the RICS HomeBuyer Service

E Outside of the property (continued)

E6 Outside doors (including patio doors)

Front Door
The external door to the front of the home is timber framed with glazed side panels.
Condition rating 1.
The glazing to the door is unlikely to be safety glass and will be a safety hazard for anyone who falls against it. You should resolve this now (see section J3).

Rear door
The external door to the rear of the home is PVC framed with glazed panels. **Condition rating 1.**

You should ask your legal adviser to check whether this door has either Building Regulation approval or has been installed by a contractor registered with FENSA (see section I1). This is a government-approved trade association whose members can self-certify that their installations meet the standards of the Building Regulations. If it was installed before April 2002 or does not have either of these, you should ask an appropriately qualified person to assess the quality of the door.

E7 Conservatory and porches

There is an open porch to the front entrance. **Condition rating 1.**

E8 Other joinery and finishes

The external joinery includes fascias, soffits, bargeboards, and timber boarding. These are made of timber and the surfaces are painted. The fascia around the main roof and rear flat roof are partly rotten. The decorations to most surfaces are beginning to flake off. **Condition rating 2.**

Without a protective decorative finish, timber will rot, requiring costly repairs. To prevent this, you should redecorate the external surfaces soon. Before decoration, most external surfaces need preparation and some repair. Although this is normal, this repair work can increase the cost of the work. To redecorate external surfaces safely, contractors will have to use appropriate access equipment (e.g. scaffolding, hydraulic platforms, etc.). This can increase the cost of the work.

E9 Other

Not applicable

 Please read the 'Description of the RICS HomeBuyer Service' (inside back cover) for details of what is, or is not, inspected.

© RICS 2009

Property Address: 52 Recession Avenue, Coventry, Warwickshire, CV0 8JJ
Serial Number: 43f6202a-7431-45a0-a819-c86a909ead96

 the mark of property professionalism worldwide

Inside of the property

I could not inspect the floor surfaces because they were covered with fitted floor coverings.

I could not inspect the underside of the floors on the ground floor because there were no access hatches.

I could not fully inspect the roof space because stored items restricted my inspection and the flooring in the roof space was not safe to use.

Condition rating

F1 Roof structure

Flooring in the roof space
A number of loose boards are laid across the floor of the roof space in a haphazard fashion. These are not fixed and are a safety hazhard (see J3). **Condition rating 3**

The loose boards should be removed. If you plan to use the roof space for light storage, you should provide suitable fixed flooring.

Party wall between 52 and 50 Recession Avenue
The wall between this and the neighbouring dwelling (called the party wall) is made of brick. The wall is partly missing just below the junction with the roof covering and could allow fire to pass from one property to another. **Condition rating 2.**

To prevent the spread of fire from one dwelling to another, you should build up the wall and seal any gaps to the underside of the roof covering soon. The owner of the neighbouring property may have a number of legal rights over this party wall. You should check with your legal adviser before you do any work (see section I3).

Ventilation to the roof space
The main roof space does not have any ventilation. **Condition rating 2.**

A lack of ventilation in the roof space can cause mould growth and wood rot. You should provide additional ventilation soon. To carry out repair work at high level safely, contractors will have to use appropriate access equipment (e.g. …

© RICS 2009

Property Address: 52 Recession Avenue, Coventry, Warwickshire, CV0 8JJ
Serial Number: 43f6202a-7431-45a0-a819-c86a909ead96

F Inside of the property (continued)

F1 Roof structure

... scaffolding, hydraulic platforms, etc.). This can increase the cost of the work.

Thermal insulation to the roof space
The roof space has a layer of thermal insulation on the top of the ceiling and this is less than the current standards. **Condition rating 2.**

A lack of thermal insulation will result in lower internal temperatures and higher heating costs. You should increase the insulation soon.

Structure of the roof
The roof structure is made of timber. **Condition rating 1.**

F2 Ceilings

The ceilings are made of a mixture of older plaster supported on thin wooden strips (called lath and plaster) and modern plasterboard. The ceiling in the lounge is cracked and most of the ceilings in the home are covered in a material that may contain asbestos (see section J3). **Condition rating 2.**

Lath and plaster ceilings are difficult to repair as large areas can be dislodged by the repair work itself. In some cases, it may be economic to replace the whole ceiling and this can be more costly than repairs to modern ceilings.

The textured decorative coating to the ceilings may contain small amounts of asbestos fibres. If these are disturbed, they could be a safety hazard. When you want to decorate or repair the ceilings, you will have to use a contractor experienced in this type of work or an asbestos specialist. This can be costly.

F3 Walls and partitions

The internal walls and partitions are made of bricks or a timber framework that have been plastered either side. There are a number of problems:

- The original internal wall between the front and rear lounge areas has been removed. The walls and floors around the new opening have distorted and cracked (for example, a crack to the lounge ceiling (F2) and distortion to the doors above). .
- The wall between the kitchen and rear lounge is affected by dampness at low level near the radiator in the kitchen but I could not discover the cause.

Condition rating 3 (further investigation)

Where an internal wall/part of an internal wall is removed or altered, it is important that the other parts of the dwelling be properly supported. ...

Please read the 'Description of the RICS HomeBuyer Service' (inside back cover) for details of what is, or is not, inspected.

© RICS 2009

Property Address: 52 Recession Avenue, Coventry, Warwickshire, CV0 8JJ
Serial Number: 43f6202a-7431-45a0-a819-c86a909ead96

F Inside of the property (continued)

F3 Walls and partitions

... If not, the unsupported parts can distort or, in the worst cases, collapse. I did see distortion around this opening and this indicates the alteration work is not adequate. You should ask an appropriately qualified professional to investigate whether the building is properly supported. This will involve removing parts of the floor and wall and you should discuss this with the current owner. Although the extent of the repairs will not be known until the investigations have been completed, it is likely to include strengthening work and this will be costly.

The dampness to the kitchen wall should be investigated as described in E4 above. To do this properly, parts of the property will have to be disturbed and you should discuss this with the current owner.

Tiles cover the wall surfaces in the bathroom from floor to ceiling. These have been painted and much of the paint is now peeling. You may want to replace the tiling but this may damage the plaster beneath.

F4 Floors

The floors to the ground and first floors are a mixture of solid floors finished with ceramic tiles and timber floors with floor boarding. **Condition rating 1.**

Like many older solid floors (usually before the 1940s), the floor in the kitchen may not have a barrier against dampness from the ground (called a damp-proof membrane or DPM). Although I could see no problems, the dampness that affects the adjacent internal wall may also affect the floor (see F3).

F5 Fireplaces, chimney breasts and flues

Fireplaces
The property includes two fireplaces and chimney breasts and these include an open fire in the original fireplace in the front bedroom, a fuel effect gas fire in the front lounge area. The open fire has a hearth made of materials that can burn easily. There is not enough ventilation and the gas fuel effect fire does not have a service record. These are safety hazards (see J3).
Condition rating 3 (further investigation).

For an open fire to work safely, it should have permanent ventilation to the room and if not, it could be a safety hazard for the occupants. Additionally, it also needs a non combustible hearth in front to prevent accidental fires. The open fire in the front bedroom has neither and should not be used.

I did not see evidence that the gas fuel effect fire has been properly maintained or installed (see G4).

Property Address: 52 Recession Avenue, Coventry, Warwickshire, CV0 8JJ
Serial Number: 43f6202a-7431-45a0-a819-c86a909ead96

F Inside of the property (continued)

F5 Fireplaces, chimney breasts and flues

Chimney breasts
The chimney breast has been removed from the rear lounge and rear bedrooms. The remaining chimney breast in the roof space is unsupported. **Condition rating 3 (further investigation).**

I saw evidence that this has not been done properly as the remaining parts are not adequately supported. You should ask an appropriately qualified professional to investigate whether there is a problem and recommend any repairs, This may involve removing parts of the floor and wall and you should discuss this with the property owner. This work may be costly.

The owner of the neighbouring property may have a number of legal rights over this chimney breast. You should check with your legal adviser before you do any work (see section I3).

F6 Built-in fittings (eg. built in kitchen, not appliances)

The kitchen and utility fittings are a mixture of timber framed units with panelled doors and drawers. The fittings are worn, damaged and some are incomplete. **Condition rating 2.**

Discontinued parts for drawers and doors, etc. can make repair work difficult and expensive. Therefore, you should plan for higher maintenance costs with these older fittings. Many buyers would plan to replace older kitchen fittings.

F7 Woodwork (for example staircase and joinery)

The staircase
The joinery items consist of: doors and door frames, skirtings, picture rails, and a staircase. These are made of timber. The staircase does not have a handrail or the associated vertical supports and this is a safety hazard. **Condition rating 3.**

Along the side of the staircase and around the edge of the landing, vertical pieces of wood (called balusters or spindles) should support the handrail. These are missing and should be replaced now.

Bedroom doors
The internal doors to the main front and rear bedrooms do not shut correctly because of the distortion in the room below (see F3). **Condition rating 2.**

Please read the 'Description of the RICS HomeBuyer Service' (inside back cover) for details of what is, or is not, inspected.

© RICS 2009

Property Address: 52 Recession Avenue, Coventry, Warwickshire, CV0 8JJ
Serial Number: 43f6202a-7431-45a0-a819-c86a909ead96

the mark of property professionalism worldwide

F8 Bathroom fittings

The fittings and appliances in the first floor bathroom comprise of a bath with shower over, a WC and a wash hand basin. The sealant around the edges of the fitting are poorfly applied. **Condition rating 2.**

The sealant around the edges of the sanitary fittings prevents excess water from seeping behind and affecting the adjacent surfaces. You should replace this soon.

F9 Other

None

Please read the 'Description of the RICS Homebuyer Service' (inside back cover) for details of what is, or is not, inspected.

© RICS 2009

Property Address: 52 Recession Avenue, Coventry, Warwickshire, CV0 8JJ
Serial Number: 43f6202a-7431-45a0-a819-c86a909ead96

the mark of property professionalism worldwide

A Surveyor's Guide to the RICS HomeBuyer Service

Services

RICS Home Buyer REPORT...

Services are generally hidden within the construction of the property and, as a result, only the visible parts of the available services can be inspected. The surveyor does not carry out specialist tests. The visual inspection does not assess the services to make sure they work properly and efficiently and meet modern standards.

Condition rating

G1 Electricity

Safety warning: You should have your electrical installations inspected and tested regularly to protect your home from damage and to avoid putting your safety at risk. Guidance published by the Institution of Electrical Engineers recommends that electrical installations should be inspected and tested at least every 10 years and when the occupiers of the property change. All electrical work carried out after 1 January 2005 should be recorded on an Electrical Installation Certificate.

There is a mains electricity supply and the meter and the consumer unit is located in the understairs cupboard in the ground floor hallway. The electricity supply was on when I inspected and the visible wiring is in PVC. The electrical system is not covered by a current inspection and testing certificate and the it does not have sufficient protection against electrocution and fire. The bathroom lacks essential safety features (for example, the bathroom light). These are safety hazards (see section J3). **Condition rating 3.**

Modern electrical systems usually incorporate a residual current device (RCD) and miniature circuit breakers (MCBs) that provides a higher level of protection against electrocution and fire. Although installing these is not a strict requirement, I would strongly recommend this to increase your personal safety.

Electrical installations in bath and shower rooms can be dangerous and are covered by strict regulations. The bathroom light is unlikely to meet current standards.

There is no evidence the system has been tested. You should ask an appropriately qualified person to inspect the electrical system and carry out any necessary repairs now.
Although I do not know the extent of the repair and improvement works required, you should plan for partial or complete replacement. This will be disruptive and costly.

 Please read the 'Description of the RICS Homebuyer Service' (inside back cover) for details of what is, or is not, inspected.

© RICS 2009

Property Address: 52 Recession Avenue, Coventry, Warwickshire, CV0 8JJ
Serial Number: 43f6202a-7431-45a0-a819-c86a909ead96

G2 Gas/oil

Safety warning: All gas and oil appliances and equipment should regularly be inspected, tested, maintained and serviced by a registered 'competent person' and in line with the manufacturer's instructions. This is important to make sure that the equipment is working correctly, to limit the risk of fire and carbon monoxide poisoning, and to prevent carbon dioxide and other greenhouse gases from leaking into the air. For more advice, contact the Gas Safe Register for gas installations, and OFTEC for oil installations.

There is a mains gas supply to the home. The control valve and the meter are in the understairs cupboard. This is loose and can be easily moved and is a safety hazard.
Condition rating 3 (further investigation).

This movement can damage the gas pipe connections and so is a safety hazard. You should ask an appropriately qualified person to inspect the gas installation system now.

G3 Water

There is an independent mains water supply to the property and the external stop valve is in the public pavement at the top of the side passage and the stop valve within the property is beneath the kitchen sink. The internal water pipes are in copper. **Condition rating 1.**

G4 Heating

The property is heated by a fixed gas fired central heating system consisting of a wall mounted condensing combination boiler with radiators in every room. There is a decorative fuel effect gas fire in the front lounge. There is no evidence that the boiler or the fire has been checked or serviced within the last 12 months and the drain from the boiler may not be properly installed. These could be safety hazards. **Condition rating 3.**

Heating installations should be installed and serviced regularly (usually every year) by an appropriately qualified person who is registered under the government-approved competent person scheme. The competent person will leave appropriate documentation with the homeowner that identifies the type and extent of the work done. This should include all heating systems and appliances, for example, boilers, individual room heaters, all open fires, etc. Heating systems and appliances that have not been checked may be a safety hazard. I did not see evidence that the appliances in this property have been checked, you should do this now and you should not use appliances until this has been done.

Property Address: 52 Recession Avenue, Coventry, Warwickshire, CV0 8JJ
Serial Number: 43f6202a-7431-45a0-a819-c86a909ead96

G Services (continued)

G5 Water heating

The hot water is provided by the main heating boiler that provides instantaneous hot water without a storage tank. There is no evidence that the boiler has been checked or serviced within the last 12 months and the drain from the boiler may not be properly installed. These could be safety hazards (see G4). **Condition rating 3**

G6 Drainage

Below ground drainage
The property is connected to the public sewer and the drainage system consists of:

- to the rear, a combined drain that takes surface water (e.g. rainwater, water from yard drains) and foul water (waste water from WCs, baths, showers, sinks, basins and dishwashers);
- to the front, a drain that takes surface water only.

I saw a number of problems with the inspection chambers:

- The covers to the front and rear are broken and are a safety hazard (see J3);
- The walls in the chambers are badly crumbling and cracked
- The drain pipes along the bottom are badly cracked and partially blocked.

Additionally, a large shrub outside the kitchen door may be damaging the drains and a new drainage connection for the rainwater pipe from the utility room is required (see E2). **Condition Rating 3**

You should repair the metal covers and frames now because they are a safety hazard as people can fall down into the inspection chambers. The walls/benching in the inspection chambers are in a poor condition and you should repair these now to prevent the debris blocking the drain. You may find it economic to rebuild the front inspection chamber completely.

The roots from the shrub may be getting into the drainage system. These may continue to grow and cause blockages that are difficult to clear. You should ask an appropriately qualified person to inspect this now. The repair may be costly and you may have to remove the shrub.

The new gulley and underground drainage connection for the utility room rainwater pipe will disturb the garden and the paths above it.

 Please read the 'Description of the RICS Homebuyer Service' (inside back cover) for details of what is, or is not, inspected.

© RICS 2009

Property Address: 52 Recession Avenue, Coventry, Warwickshire, CV0 8JJ
Serial Number: 43f6202a-7431-45a0-a819-c86a909ead96

Appendices

G6 Drainage

The drain from this home passes through the land belonging to the neighbours (both No. 50 and 54 Recession Avenue) and a legal agreement should cover this. You should ask your adviser to check this (see section I3).

Above ground drainage
The bathroom, toilet, and kitchen fittings are connected to the underground drains through a series of plastic pipes called a soil stack just outside the kitchen door. The plastic waste pipe from the washing machine in the utility room runs across the rear step and is a safety hazard (see J3). **Condition rating 3.**

The waste pipe from the washing machine should be redirected to the new drainage connection (see above).

G7 Common Services

None

Please read the 'Description of the RICS Homebuyer Service' (inside back cover) for details of what is, or is not, inspected.

© RICS 2009

Property Address: 52 Recession Avenue, Coventry, Warwickshire, CV0 8JJ
Serial Number: 43f6202a-7431-45a0-a819-c86a909ead96

Grounds (including shared areas for flats)

I could not inspect the inside of the garage because it was locked and I could not see through the window.

I could not get access to the rear roadway because the security gates were locked.

Condition rating

H1 Garage

The garage is in the rear part of the rear garden. It is built of pre-cast concrete with a corrugated cement sheet roof covering. **Not inspected**.

The garage roof is made of a material that may contain asbestos fibres. You should not disturb this without prior advice from a licensed asbestos specialist or a contractor experienced in this type of work. Additionally, the cement-based roof sheets are very fragile and you should take care when carrying out maintenance work.

H2 Other

None

H3 General

The boundaries
The boundaries of the home consist of:

- the rear garden - the southwest, southeast, and northeast boundaries consist of timber panel fences supported by timber posts. A number of horizontal timber support rails and fencing panels need to be replaced;
- the front garden - the south west and southeast boundaries consist of timber fencing panels supported by timber posts that are in a poor condition. The northwest boundary consists of a dwarf masonry wall and a portion of this has collapsed. This needs to be repaired or replaced.

The legal documents that describe the ownership of the home (the deeds) usually describe which of the neighbouring owners are responsible for the upkeep of the boundaries. You should ask your legal adviser to advise you on this (see section I3).

 Please read the 'Description of the RICS Homebuyer Service' (inside back cover) for details of what is, or is not, inspected.

© RICS 2009

Property Address: 52 Recession Avenue, Coventry, Warwickshire, CV0 8JJ
Serial Number: 43f6202a-7431-45a0-a819-c86a909ead96

 the mark of property professionalism worldwide

H Grounds (continued)

Footpaths
The main footpaths include a concrete paving slab path to the front garden and a patio at the rear. A number of the slabs are loose and are uneven and are safety hazards (see J3).

The grounds

- the front garden - there is a row of evergreen shrubs just behind the dwarf masonry wall. These could undermine the wall and reduce the daylight to the house. You should reduce their height or remove them completely.
- the rear garden - this is very overgrown with a number of large shrubs and small trees. You should consider reducing their size or removing them altogether to prevent future damage to paths, fences and drains.

Rear access road
I did not inspect this because the gates were locked.

Issues for your legal advisers

The surveyor does not act as 'the legal adviser'. However, if, during the inspection, the surveyor identifies issues that the legal advisers may need to investigate further, the surveyor will refer to these in the report. The surveyor will not comment on any legal documents or any materials that may be included in a Home Information Pack.

I1 Regulation

You should ask your legal adviser to confirm whether the roof covering to the main roof (E1), replacement windows (E5), door (E6) and removal of internal wall in the lounge (F2) have received Building Regulation approval (including the issuing of a final completion certificate) from the local council and advise on the implications.

I2 Guarantees

You should ask your legal adviser to confirm whether the walls (especially the rear wall - see E4) that have been injected with the chemical damp-proof course (DPC) are covered by a guarantee or warranty and advise on the implications.

You should ask your legal adviser to confirm whether the replacement windows and the rear door are covered by a guarantee or warranty and advise on the implications.

I3 Other matters

Part of the home is shared with the neighbouring owner (the chimney, E1, the party wall in the roof space, F1, and the rear chimney breast, F5). Before you carry out any repairs or alterations, you may have to get their agreement to the work. You should ask your legal adviser to confirm this and explain the implications.

The drains of this home pass under the adjacent property of both 50 and 54 Recession Avenue (see section G6). You should ask your legal adviser to confirm this and explain the implications. In some cases, the costs of the maintenance and repair of the whole drain may be shared with the neighbouring owners.

Access to the rear garage is by a private road that may not be maintained by the local council. Security gates prevent access to this road. Your legal adviser should confirm this and advise you about the implications. The responsibility for private roads can be shared between the people that use it and this can be costly especially if the road is in poor condition.

Property Address: 52 Recession Avenue, Coventry, Warwickshire, CV0 8JJ
Serial Number: 43f6202a-7431-45a0-a819-c86a909ead96

J Risks

This section covers defects that need repairing or replacing, as well as issues that have existed for a long time and cannot reasonably be changed but may present a health and safety risk or hazard. If the defects affect specific elements, they will be referred to in the individual section. Other risks or hazards that also affect value are also reported.

J1 Risks to the building

Structural movement
The home is affected by building movement caused by the removal of the internal partition in the lounge (see F3) and the removal of the chimney breasts (F5).

Dampness
The home is affected by dampness to the rear lounge (E4) and the kitchen (F3) but I could not identify the cause.

Timber defects
None

J2 Risks to the grounds

Contamination
None.

Flooding
None.

J3 Risks to people

Asbestos
The following issues are safety hazards:

- E5 - lack of escape from window of small front bedroom
- E6 - front door, lack of safety glass
- F1 - loose floorboards
- F2 - textured coatings to ceilings that may contain asbestos
- F7 - missing handrail and balusters to staircase

J Risks (continued)

J3 Risks to people

- G1 - unsafe electrical system
- G2 - poorly secured gas meter
- G4 - lack of service or inspection of boiler and gas fire
- G6 - broken inspection chamber covers
- H1 - possible asbestos in garage roof
- H3 - tripping hazards to path and patio

Health and safety advice
None.

J4 Other

None

Please read the 'Description of the RICS HomeBuyer Service' (inside back cover) for details of what is, or is not, inspected.

© RICS 2009

the mark of property professionalism worldwide

Property Address: 52 Recession Avenue, Coventry, Warwickshire, CV0 8JJ
Serial Number: 43f6202a-7431-45a0-a819-c86a909ead96

K Valuation

RICS Home Buyer REPORT...

In arriving at my valuation, I made the following assumptions.

With regard to the materials, construction, services, fixtures and fittings, and so on I have assumed that:

- an inspection of those parts that I could not inspect would not identify significant defects or a cause to alter the valuation;
- no dangerous or damaging materials or building techniques have been used in the property;
- there is no contamination in or from the ground, and the ground has not been used as landfill;
- the property is connected to, and has the right to use, the mains services mentioned in the report; and
- the valuation does not take account of any furnishings, removable fittings or sales incentives.

With regard to legal matters I have assumed that:

- the property is sold with 'vacant possession' (your legal adviser can give you more information on this term);
- the condition of the property, or the purpose the property is, or will be, used for does not break any laws;
- no particularly troublesome or unusual restrictions apply to the property, that the property is not affected by problems which would be revealed by the usual legal inquiries, and that you have applied for and acted in line with all necessary planning permission and Building Regulations permission (including permission to make alterations); and
- the property has the right to use the mains services on normal terms, and that the sewers, mains services and roads giving access to the property have been 'adopted' (that is, they are under local-authority, not private, control).

Any additional assumptions relating to the valuation:

In my opinion the current market value on
25/12/2009
as inspected was:

£ 135,000

(amount in words)
One hundred and thirty five thousand pounds

TENURE
Freehold

AREA OF PROPERTY (M^2)
94

In my opinion, the current reinstatement cost of the property (see below) is

£ 125,000

* Approximate gross external / internal area of the building or flat.

the mark of property professionalism worldwide

Property Address: 52 Recession Avenue, Coventry, Warwickshire, CV0 8JJ
Serial Number: 43f6202a-7431-45a0-a819-c86a909ead96

A Surveyor's Guide to the RICS HomeBuyer Service

K Valuation (continued)

Your legal advisers, and other people who carry out property conveyancing, should be familiar with these assumptions and are responsible for checking those concerning legal matters.

My opinion of the market value shown here could be affected by the outcome of the enquiries by your legal advisers (section I) and/or any further investigations and quotations for repairs or replacements. The valuation assumes that your legal advisers will receive satisfactory replies to their enquiries about any assumptions in this report.

Other considerations affecting value:

RICS Home Buyer REPORT...

You can find information about the assumptions I have made in calculating this reinstatement cost in the 'Description of the RICS HomeBuyer Service' provided. The reinstatement cost is the cost of rebuilding an average home of the type and style inspected to its existing standard using modern materials and techniques, and by acting in line with current Building Regulations and other legal requirements. This will help you decide on the amount of buildings insurance cover you will need for the property.

the mark of property professionalism worldwide

Property Address: 52 Recession Avenue, Coventry, Warwickshire, CV0 8JJ
Serial Number: 43f6202a-7431-45a0-a819-c86a909ead96

Appendices

RICS
Home Buyer
REPORT...

Surveyor's declaration

"I confirm that I have inspected the property and prepared this report, and the market value given in the report."

Signature:

Surveyor's name:	Phil Parnham
Surveyor's RICS number:	007
Qualifications:	BSc, MRICS
Street:	BlueBox House Residential Way Coventry
Town:	Coventry
Postcode:	CV1 0AA
E-mail address:	MrParnham@blueboxpartners.com
Website:	http://www.ricssurveywriter.com
Phone number:	1234 5678
Fax number:	

Date this report was produced:	26/12/2009
Property address:	52 Recession Avenue, Coventry, Warwickshire, CV0 8JJ
Customer's name:	Mr A.N. Other

RICS Disclaimer

This document is issued in blank form by the Royal Institution of Chartered Surveyors (RICS) and is available only to parties who have signed a licence agreement with RICS.

RICS gives no representations or warranties, express or implied and no responsibility or liability is accepted for the accuracy or completeness of the information inserted in the document or any other written or oral information given to any interested party or its advisers. Any such liability is expressly disclaimed.

the mark of property professionalism worldwide

Property Address: 52 Recession Avenue, Coventry, Warwickshire, CV0 8JJ
Serial Number: 43f6202a-7431-45a0-a819-c86a909ead96

What to do now

Getting estimates

The cost of repairs may influence the amount you are prepared to pay for the property. Before you make a legal commitment to buy the property, you should get reports and estimates for all the repairs and further investigations the surveyor may have identified.

You should get at least two estimates from experienced contractors who are properly insured.

You should also:

- ask them for references from people they have worked for;
- describe in writing exactly what you want them to do; and
- get the contractors to put the estimates in writing.

Some repairs will need contractors with specialist skills and who are members of regulated organisations (for example, electricians, gas engineers, plumbers, and so on). Some work may also need you to get Building Regulations permission or planning permission from your local authority. You can get more advice on these from the Government's website at www.direct.gov.uk/en/HomeAndCommunity/Planning/index.htm.

Further investigations

If the surveyor was concerned about the condition of a hidden part of the building or could only see part of a defect, or does not have the specialist knowledge to fully assess part of the property, they may have recommended that further investigations should be carried out to discover the true extent of the problem.

Who you should use for these further investigations

You should ask an appropriately qualified person. It is not possible to tell you which one because specialists belonging to several different types of organisations will be able to do this. For example, qualified electricians can belong to five different government-approved schemes. If you want further advice, please contact the surveyor. You can also get more advice on how to find an appropriately qualified person from the Government's website mentioned above.

What will the further investigations involve?

This will depend on the type of problem but, to do this properly, parts of the home may have to be disturbed and so you should discuss this matter with the current owner. In some cases, the cost of the investigation may be high.

When to do the work

The condition ratings help describe the urgency of the repair and replacement work. The following summary may help you decide when to do the work.

- Condition rating 2 - repairs should be done soon. Exactly when will depend on the type of problem but it usually does not have to be done right away. Many repairs could wait weeks or months, giving you time to organise suitable reports and estimates.

- Condition rating 3 - repairs should be done as soon as possible. The speed of your response will depend on the nature of the problem. For example, repairs to a badly leaking roof or a dangerous gas boiler need to be carried out within a matter of hours, while other less important critical repairs could wait for a few days.

Warning

Although repairs of elements with a condition rating 2 are not considered urgent, if they are not dealt with they may develop into more serious repairs. Flat roofs and gutters are typical examples. These can quickly get worse without warning and result in serious leaks.

As a result, you should regularly check elements with a condition rating 2 to make sure they are not getting worse.

Property Address: 52 Recession Avenue, Coventry, Warwickshire, CV0 8JJ
Serial Number: 43f6202a-7431-45a0-a819-c86a909ead96

Description of the RICS HomeBuyer Service

The service

The RICS HomeBuyer Service includes:
- an inspection of the property (see 'The inspection' below);
- report based on the inspection (see 'The report' below); and
- a valuation, which is part of the report (see 'The valuation' below).

The surveyor who provides the RICS HomeBuyer Service aims to give you professional advice to help you to:
- make a reasoned and informed decision on whether to go ahead with buying the property;
- make an informed decision on what is a reasonable price to pay for the property;
- take account of any repairs or replacements the property needs; and
- consider what further advice you should take before exchanging contracts (if the property is in England, Wales, Northern Ireland, the Isle of Man or the Channel Islands) or concluding an offer (if the property is in Scotland).

The inspection

The surveyor inspects the inside and outside of the main building and all permanent outbuildings, but does not force or open up the fabric. This means that they do not take up carpets, floor coverings or floorboards, move furniture, remove the contents of cupboards, remove secured panels or undo electrical fittings. If necessary, the surveyor will carry out parts of the inspection when standing at ground level from public property next door.

The surveyor may use equipment such as a damp-meter, binoculars and torch, and may use a ladder for flat roofs and for hatches no more than three metres above level ground (outside) or floor surfaces (inside) if it is safe to do so.

Services to the property

Services are often hidden within the construction of the property and, as a result, only the visible parts of the available services can be inspected. The surveyor will not carry out specialist tests, or test or assess the efficiency of electrical, gas, plumbing, heating or drainage installations (or whether they meet current regulations) or the inside condition of any chimney, boiler or other flue.

Outside the property

The surveyor inspects the condition of boundary walls, fences, permanent outbuildings and areas in common (shared) use. To inspect these areas, the surveyor walks around the grounds and any neighbouring public property. Buildings with swimming pools and sports facilities are also treated as permanent outbuildings, but the surveyor will not report on the leisure facilities, such as the pool itself and its equipment, landscaping and other facilities (for example, tennis courts and temporary outbuildings).

Flats

When inspecting flats, the surveyor will assess the general condition of outside surfaces of the building, as well as its access areas (shared hallways, staircases and so on). The surveyor will inspect roof spaces only if they can gain access to them from within the property. The surveyor will not inspect drains, lifts, fire alarms and security systems.

Dangerous materials, contamination and environmental issues.

The surveyor will not make any enquiries about contamination or other environmental dangers. However, if they suspect a problem, they should recommend a further investigation.

The surveyor may assume that no harmful or dangerous materials have been used in the construction, and does not have a duty to justify making this assumption. However, if the inspection shows that these materials have been used, the surveyor must report this and ask for further instructions.

The surveyor will not carry out an asbestos inspection, and will not act as an asbestos inspector when inspecting properties that may fall within the Control of Asbestos Regulations 2006. With flats, the surveyor will assume that there is a dutyholder (as defined in the regulations), and that an asbestos register and an effective management plan is in place which does not need any immediate payment or present a significant risk to health. The surveyor will not consult the dutyholder.

The report

The surveyor produces a report of their inspection for you to use, but cannot accept any liability if it is used by anyone else. If you decide not to act on the advice in the report, you do this at your own risk. The report focuses on matters that, in the surveyor's opinion, may affect the value of the property if they are not dealt with.

The report is in a standard format and includes the following sections.

A Introduction to the report
B About the inspection
C Summary of the condition ratings
D About the property
E Outside of the property
F Inside of the property
G Services
H Grounds (including shared areas for flats)
I Issues for your legal advisers
J Risks
K Valuation
L Surveyor's declaration
What to do now
Description of the HomeBuyer Service
House diagram

The surveyor gives condition ratings to the main parts (or 'elements') of the main building, garage, and some outside elements. The condition ratings are described as follows.

Condition rating 1
No repair is currently needed. The property must be maintained in the normal way.

Condition rating 2
Defects that need repairing or replacing but are not considered to be either serious or urgent. The property must be maintained in the normal way.

Condition rating 3
Defects that are serious and/or need to be repaired, replaced or investigated urgently.

NI Not inspected

The surveyor will note in their report if they were not able to check any parts of the property that the inspection would normally cover. If the surveyor is concerned about these parts, the report tells you about any further investigations that are needed.

The surveyor will not report on the cost of any work to put right defects or make recommendations on how these repairs should be carried out. However, there is general advice in the 'What to do now' section at the end of the report.

If, during the inspection, the surveyor identifies issues that your legal advisers may need to investigate further, the surveyor will refer to these in the report. The surveyor will state the Energy Efficiency Rating and Environmental Impact Rating as stated on the Energy Performance Certificate on the home, if available. The surveyor will not specifically comment on any legal documents or any documents that may be included in a Home Information Pack or Home Report.

The surveyor reports on property-related risks or hazards that will include defects that need repairing or replacing, as well as issues that have existed for a long time and cannot reasonably be changed but may present a health and safety risk or hazard.

If the property is leasehold, the surveyor will give you general advice and details of questions you should ask your legal advisers.

The valuation

The surveyor gives their opinion on both the market value of the property and the reinstatement cost at the time of their inspection. (See the 'Reinstatement cost' section opposite.)

Market value

'Market value' is the estimated amount a property should exchange for, on the date of the valuation, between a willing buyer and a willing seller, in an arm's length transaction after the property was properly marketed wherein the parties had each acted knowledgeably, prudently and without compulsion.

When deciding on the market value, the surveyor also makes the following assumptions.

The materials, construction, services, fixtures and fittings, and so on

The surveyor will assume that:
- an inspection of those parts which have not yet been inspected would not identify significant defects or cause the surveyor to alter their valuation;
- no dangerous or damaging materials or building techniques have been used in the property;
- there is no contamination in or from the ground, and the ground has not been used as landfill;
- the property is connected to, and has the right to use, the mains services mentioned in the report; and
- the valuation does not take account of any furnishings, removable fittings and sales incentives of any description.

Legal matters

The surveyor will assume that:
- the property is sold with 'vacant possession'(your legal adviser can give you more information on this term);
- the condition of the property, or the purpose that the property is, or will be, used for does not break any laws;
- no particularly troublesome or unusual restrictions apply to the property, that the property is not affected by problems which would be revealed by the usual legal enquiries, and that you have applied for and acted in line with all necessary planning permission and Building Regulations permission (including permission to make alterations); and
- the property has the right to use the mains services on normal terms, and that the sewers, mains services and roads giving access to the property have been 'adopted' (that is, they are under local-authority, not private, control).

The surveyor will report any more assumptions they have made or found not to apply.

If the property is leasehold, the general advice referred to above will explain what other assumptions the surveyor has made.

Reinstatement cost

'Reinstatement cost' is the cost of rebuilding an average home of the type and style inspected to its existing standard using modern materials and techniques and in line with current Building Regulations and other legal requirements.

This includes the cost of rebuilding any garage, boundary or retaining walls and permanent outbuildings, and clearing the site. It also includes professional fees, but does not include VAT (except on fees).

The reinstatement cost will help you decide on the amount of buildings insurance cover you will need for the property.

Standard terms of engagement

1 The service - the surveyor will provide the standard RICS HomeBuyer Service ('the service') described in the 'Description of the RICS HomeBuyer Service', unless you and the surveyor agree in writing before the inspection that the surveyor will give you additional advice.

2 The surveyor providing the service will be a full Member or Fellow of the Royal Institution of Chartered Surveyors, who has the skills, knowledge and experience to survey, value and report on the property.

3 Before the inspection - you will tell the surveyor if there is already an agreed, or proposed, price for the property, and if you have any particular concerns (such as plans for extension) about the property.

4 Terms of payment - you agree to pay the surveyor's fee and any other charges agreed in writing.

5 Cancelling this contract - you are entitled to cancel this contract by giving notice to the surveyor's office at any time before the day of the inspection. The surveyor will not provide the service (and will report this to you as soon as possible) if, after arriving at the property, they decide that:

a they lack enough specialist knowledge of the method of construction used to build the property; or
b it would be in your best interests to have a building survey and a valuation, rather than the RICS HomeBuyer Service.

If you cancel this contract, the surveyor will refund any money you have paid for the service, except for any reasonable expenses. If the surveyor cancels this contract, they will explain the reason to you.

6 Liability - the report is provided for your use, and the surveyor cannot accept responsibility if it used by anyone else.

Complaints handling procedure

The surveyor will have a complaints handling procedure and will give you a copy if you ask.

Note: These terms form part of the contract between you and the surveyor.

A Surveyor's Guide to the RICS HomeBuyer Service

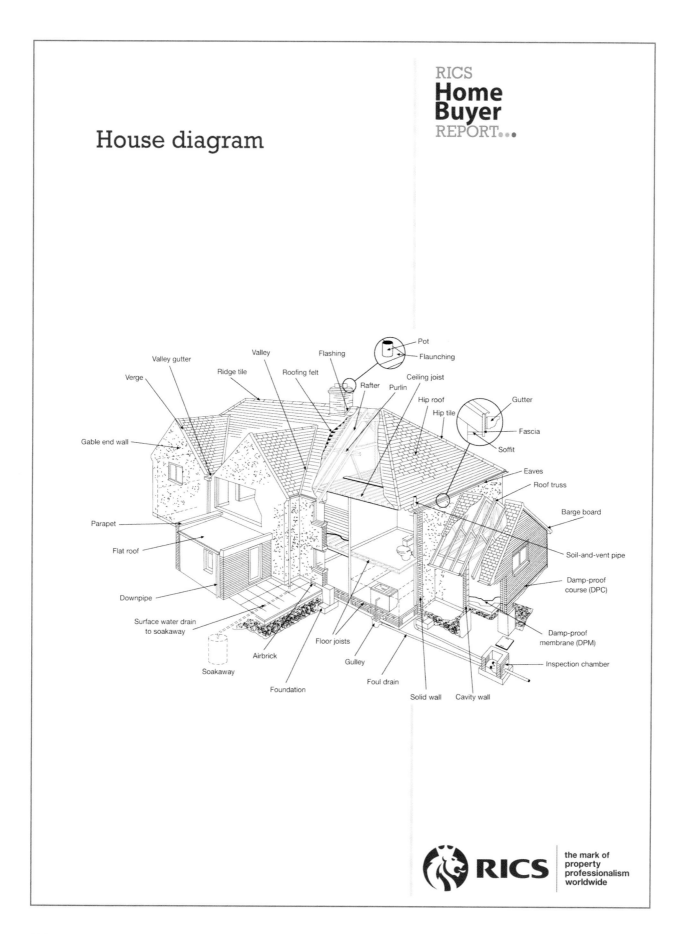

Index

asbestos
   condition rating system
      lagging to flue 62–63
      soffit panels 63

bathroom fittings
   new element, condition rating 67–68
binoculars 13
British Geological Society 11
   radon map 11
Building Research Establishment (BRE)
   radon advice 11
building services
   condition rating of heating systems 58–59
   special status of further investigations, 'not specified' and condition rating 41
      annual servicing and other safety checks 42
      change in culture 43
      changing regulatory environment 42
      impact on HIP 43
      increasing concern over safety issues 42
      standard advice 41–42

Carlsberg 'Review of Residential Property' 3
chimney stacks
   report writing
      legal issues 87
      stone built stock: example report 85
Coal Authority 10
   mining reports 10
communication
   with clients, new practice notes 5–6
compass 14
condensation
   condition rating 68, 69–70
Condition Rating Protocol 45–46
   broader issues affecting decisions 46
   building elements in flats 77
      elements exclusive to the inspected flat 77
      elements shared with block 77–78
      external shared areas 79
      services 78–79
   built-in elements 67
      bathroom fittings, new element 67–68
   case studies 46
      heating systems 53–59
      roof coverings 47–52
   conservatories and porches
      new building 66–67
      timber 'sun room' 67
      uPVC conservatory 66–67
   decorations 66
   external features including trees 74
   factors affecting value of property but not condition rated 75–76
   garages 70
      integral garage 72
      1940s garage 70–71
      1980s attached garage 71–72
   health and safety 59–60
      asbestos lagging to flue 62–63
      asbestos soffit panels 63
      gardens ponds 62
      lack of safety glazing 61
      lack of windows for emergency escape 61–62
      lead pipes 64
      objective method for assessing hazards 60–65
      serious and significant tripping hazards 64–65
      unsafe parts of building 65
   insulation, ventilation and condensation 68–69
   market value 76–77
   multiple building elements 65–66
   outbuildings 70, 73
   'risks to the building' 75
   two-stage process 46
condition ratings 37
   building services *see* building services
   deconstructing definitions 37–38
   defect 38–39
   further investigations 40
      roof space inspection 40–41
   introduction 37
   normal maintenance 39
   protocol *see* Condition Rating Protocol
   serious defects 39–40
   urgent defects 40
conservation areas
   inspection
      knowing your limitations 8
      pre-inspection checks 12
conservatories and porches
   condition rating
      new building 66–67
      timber 'sun room' 67
      uPVC conservatory 66–67
   report writing: legal issues 87
contaminated land
   factor affecting value of property but not condition rated 75

dangerous dogs
   dealing with 21
dangerous electrics
   market value and condition rating 76
decorations
   condition rating 66
defects *see* condition ratings
Department for Environment, Food and Rural Affairs (Defra)
   noise maps 9
derelict properties
   inspection, safety issues 23–24
digital cameras 14
dogs
   dangerous, dealing with 21

electrical fittings
   dangerous, market value and condition rating 76

electro-magnetic fields
    factor affecting value of property but not condition rated 75
electronic moisture meter 13
employers
    health and safety legislation 16
empty properties
    inspection, safety issues 23–24
Environment Agency 8–9
    flooding maps 9
    groundwater protection zones 9
    landfill 9
    pollution 9
environmental risk assessment
    free services 11–12

flats
    condition rating building elements 77
        elements exclusive to the inspected flat 77
        elements shared with block 77–78
        external shared areas 79
        services 78–79
    inspection 30
        extent 30–32
flight paths *see* major nuisances
flooding
    factor affecting value of property but not condition rated 75

garages
    condition rating 70
        integral garage 72
        1940s garage 70–71
        1980s attached garage 71–72
garden ponds
    condition rating system 62
gas boilers *see* heating systems
glazing
    condition rating system
        lack of safety glazing 61
        lack of windows for emergency escape 61–62
guarantees and warranties
    legal advice 87–88

health and safety 15
    condition rating system 59–60
        asbestos lagging to flue 62–63
        asbestos soffit panels 63
        gardens ponds 62
        lack of safety glazing 61
        lack of windows for emergency escape 61–62
        lead pipes 64
        objective method for assessing hazards 60–65
        serious and significant tripping hazards 64–65
        unsafe parts of building 65
    home working 19
    legislation 15–16
        employers 16
    office environment 18
    property-specific safety issues 22
        derelict and empty properties 23–24
        loft inspections 22–23
    risk assessments *see* risk assessments
    travel and general vehicle safety 18–19
        driving distractions 19
health and safety equipment 14–15
Health Protection Agency (HPA) 11
    radon 11
heating systems
    condition rating case studies 53–59
    report-writing example 86

HomeBuyer Report
    additional advice – 'up sell' your services 6
    building for future 91
    expanding market share 89
        marketing 89–90
        RICS support 90
    need for change 3–4
    preparing for inspection *see* inspection
    sample of completed report 193–224
HomeBuyer Report (3rd edition) 4–6
    new practice notes 4
        communication with client 5–6
        HSV 'additional advice' provision 6
        HSV requirements 6
        relationship with HIP 5
Homebuyer Survey and Valuation Service (HSV) 3–4
    'additional advice' provision 6
    requirements 6
Home Information Packs (HIPs)
    launch 3
    Property Information Questionnaire (PIQ) 43
    relationship with new practice notes 5
home working
    health and safety, employer's duties 19

inspection
    health and safety *see* health and safety
    knowing your limitations 7–8
        historic/traditionally constructed buildings, listing and conservation areas 8
        particular technical skills and experience 7
        prefabricated properties 7
        pre-Victorian properties 7
    liaising with occupiers 24
        building relationship with seller – on arrival 25
        dealing with seller enquiries 25
        using seller questionnaires 25, 161–163
    pre-inspection checks 8
        Defra noise maps 9
        Environment Agency 8–9
        free environmental risk assessment services 11–12
        geological information 9–11
        Health Protection Agency (HPA) 11
        listed buildings and conservation areas 12
        OFCOM (mobile phone base stations) 11–12
        regional constructional techniques and local knowledge 12
    procedure *see* inspection procedure
    risk assessments *see* risk assessments
    tools and equipment 12–13
        binoculars 13
        compass 14
        digital camera 14
        electronic moisture meter 13
        health and safety equipment 14–15
        ladder 13
        lifting equipment/crow bar 13
        measuring devices 13
        plumb bob 14
        pocket mirror 14
        protective clothing 14
        spirit level 14
        torch 13
inspection procedure 32
    inspection of flats 30
        extent 30–32
    site notes 32–34
        packages *see* site notes packages
        sample 167–192
        themes discussed at training seminars 27–30

insulation
    condition rating 68

ladders 13
landfill
    factor affecting value of property but not condition rated 75
lead pipes
    condition rating system 64
lifting equipment/crow bars 13
listed buildings
    inspection
        knowing your limitations 8
        pre-inspection checks 12
local construction techniques 12
loft inspections
    condition ratings 40–41
    safety issues 22–23

major nuisances
    factor affecting value of property but not condition rated 75
major roads *see* major nuisances
market value
    condition rating 76
        dangerous electrics 76
        major structural movement 76
        overhead pylons 76
    market valuation analysis 165
measuring devices 13
mining activities *see also* Coal Authority
    factor affecting value of property but not condition rated 75
mobile phone base stations 11

nuisances *see* major nuisances

OFCOM
    mobile phone base stations 11
office-based work
    health and safety requirements 18
outbuildings
    condition rating 70, 73
overhead pylons
    market value and condition rating 77

plumb bobs 14
pocket mirrors 14
porches *see* conservatories and porches
prefabricated properties
    inspection, knowing your limitations 7
pre-Victorian properties
    inspection, knowing your limitations 7
protective clothing 14

radon
    factor affecting value of property but not condition rated 75
    HPA information 11
railway lines *see* major nuisances
regional constructional techniques 12
report writing 81–83
    completing comment boxes 83
        advisory section 83–84
        descriptive section 83

completing Section 1: issues for legal advisers 86–87
    guarantees and warranties 87–88
cosmetic and minor items 84–85
examples 85
    gas boiler 86
    slate covered roof 86
    stone built chimney stack 85
handling technical terms 82
length of report 84
order of text 84
overall opinion 88
reporting further investigations 84
sample report 86
risk assessments
    dealing with dangerous dogs 21
    free environmental risk assessment services 11–12
    health and safety in occupied properties 20–22
    on arrival at property 20
    pre-inspection health and safety preparation 19–20
    principles 16–18
    problems posed by building 20
'risks to the building'
    condition rating 75
roof coverings
    condition rating
        case studies 47–52
        unsafe parts of building 65
    slate covered roof: report-writing example 86
roof space inspections *see* loft inspections

Single Survey in Scotland 3
site notes 32–34
    sample 167–192
site notes packages 34
    dictated site observations with handwritten sketch plans and separate checklists 35, 147–160
    handwritten site observations
        elemental structure 34–35, 95–120
        room-by-room supplement 35, 121–146
Society for the Protection of Ancient Buildings (SPAB) 12
spirit levels 14
structural movement
    market value and condition rating 76

torches 13
traditionally constructed buildings
    inspection, knowing your limitations 8
trees
    condition rating 74
tripping hazards
    serious and significant, condition rating system 64–65

ventilation
    condition rating 68, 69

warranties *see* guarantees and warranties
windows *see* glazing
work-related travel
    driving distractions 19
    travel and general vehicle safety 18–19